2019 35th Semiconductor Thermal Measurement, Modeling and Management Symposium (SEMI-THERM 2019)

San Jose, California, USA
18-22 March 2019

IEEE Catalog Number: CFP19SEM-POD
ISBN: 978-1-7281-9129-4

Copyright © 2019, SEMI-THERM Educational Foundation (STEF)
All Rights Reserved

*** *This is a print representation of what appears in the IEEE Digital Library. Some format issues inherent in the e-media version may also appear in this print version.*

IEEE Catalog Number: CFP19SEM-POD
ISBN (Print-On-Demand):   978-1-7281-9129-4
ISBN (Online):   978-1-7355325-0-9
ISSN: 1065-2221

Additional Copies of This Publication Are Available From:

Curran Associates, Inc
57 Morehouse Lane
Red Hook, NY 12571 USA
Phone: (845) 758-0400
Fax: (845) 758-2633
E-mail: curran@proceedings.com
Web: www.proceedings.com

2019 35th Semiconductor Thermal Measurement, Modeling and Management Symposium (SEMI-THERM 2019)

San Jose, California, USA
18-22 March 2019

IEEE Catalog Number: CFP19SEM-POD
ISBN: 978-1-7281-9129-4

Thirty-Fourth Annual

SEMICONDUCTOR THERMAL MEASUREMENT AND MANAGEMENT SYMPOSIUM

PROCEEDINGS 2019

San Jose, CA   USA
March 18-22, 2019

Thirty Fifth Annual

SEMICONDUCTOR THERMAL MEASUREMENT, MODELING AND MANAGEMENT SYMPOSIUM

PROCEEDINGS 2019

San Jose, CA   USA
March 18-22, 2019

The 2019 Semiconductor Thermal Measurement, Modeling and Management (SEMI-THERM) Symposium is an annual international forum for the presentation of new developments in and applications relating to generation and removal of heat within semiconductor devices, and measurement of junction temperatures under various application and environmental conditions.

Attendance at the Symposium is limited, to preserve the close interaction among attendees and presenters. The format of the symposium this year couples eight sessions of selected technical papers, a more-intimate Poster Session for one-on-one discussion of results, a luncheon talk, and a series of Workshops focused on products and techniques and an embedded tutorial.

This year, the Symposium is preceded by four Short Courses: **"Statistical Analysis Methods for Dealing with Uncertainty in Thermal Testing," "Design and Optimization of Heat Sinks," "A Holistic Approach to Improve Mission Critical Facility Performance," "Thermal Challenges in Automotive Electronics," "Introduction to the Design and Implementation of Indirect Liquid Cooling for Electronic Systems,"** and **"Design of Experiments (DOE) for Thermal Engineering."** In addition, an exhibits area offers displays of equipment, software, and other resources within the thermal measurements field.

We trust you will take advantage of the rich array of information and experiences developed by this year's Steering and Program Committees, and consider submitting an abstract for next year's SEMI-THERM.

General Chair
Adriana Rangel,
  Cisco Systems

Program Chair
Pablo Hidalgo,
  Aavid

Vice Program Chair
Marcelo del Valle,
  Intel

Symposium/Exhibitor Management/Registration
Denise Rael,
  +1- 480 839-8988,
  drael@semi-therm.org

Europe Liaisons
Clemens J. M. Lasance,
  SomeLikeItCool
John Parry,
  Mentor Graphics

Asia Liaison
Prof. Hsiao-Kang Ma,
  National Taiwan University
Wataru Nakayama,
  ThermTech International
Winston Zhang,
  Novark

Proceedings
Paul Wesling

SEMI-THERM® 35

Welcome to SEMI-THERM 35!

Adrianna Rangel
Symposium General Chair

Dear Colleagues,

Welcome to the SEMI-THERM 35 annual conference. This year, the program committee has assembled an excellent program dedicated to thermal design, thermal management as well as measurement of semiconductor systems and components.

The SEMI-THERM conference is committed to provide a forum for discussion of the latest advances in electronic thermal management. This year's program provides 6 short courses which are included with the conference registration. They are presented by experts in a variety of topics. The short courses are designed to educate our engineering community and to start our conference with an opportunity to learn and network on the first day, Monday March 18th. An evening tutorial will be presented on Tuesday night and How -to courses complete the evening on Wednesday after the exhibitor reception.

The program committee has also put together a fantastic morning program that includes technical papers, a keynote speaker and 2 luncheon speakers. On Tuesday and Wednesday afternoon, the SEMI-THERM exhibits are open to all as well as vendor workshops, with both providing opportunities to learn more about innovative thermal products and vendors.

At the Thursday luncheon, as part of SEMI-THERM tradition, an award session will include the presentation of the THERMI award, the Harvey Rosten awards, as well as the Thermal Hall of Fame, Lifetime Achievement award. The Best Paper in different categories will also be recognized.

The Harvey Rosten award recipients are James W. VanGilder, Christopher M. Healey, Michael Condor, Wei Tian, Quentin Menusier. The THERMI award recipient is Dr. Peter Raad from Southern Methodist University. The Thermal Hall of Fame Lifetime Achievement will be presented to Márta Rencz, in recognition of significant contributions to the field of electronics thermal management.

I would like to congratulate the Program Chair, Pablo Hidalgo, and Vice- Chair, Marcelo del Valle, for this year's outstanding conference program. I hope you will enjoy attending this symposium, and I look forward to a wonderful week of learning and networking.

Sincerely,

Adriana Rangel
Symposium General Chair

SEMI-THERM 35

SEMI-THERM 35 SYMPOSIUM PERSONNEL

General Chair:
Adriana Rangel, Cisco Systems adromero@cisco.com

Program Chair:
Pablo Hidalgo, Aavid
Pablo.Hidalgo@boydcorp.com

Program Vice Chair:
Marcelo del Valle, Intel marcelo.del.valle@intel.com

International Liaisons:
Clemens Lasance, Thermal Management Consultant
SomeLikeItCool lasance@onsnet.nu

John Parry, Mentor, A Siemens Business
john_parry@mentor.com

Hsiao-Kang Ma, National Taiwan University, Taiwan
skma@ntu.edu.tw

Wataru Nakayama, ThermTech International, Japan
watnakayama@aol.com

Winston Zhang, Novark, China
winstonzhang@novark.com.cn

Social Media:
Robin Bornoff Mentor, A Siemens Business
Robin_Bornoff@mentor.com

Symposium Management:
SEMI-THERM Symposium Manager
Bonnie Crystall, C/S Communications, Inc.
cscomm@earthlink.net

Proceedings IEEE Region 6:
Paul Wesling p.wesling@ieee.org

SEMI-THERM 35 Steering/Technical Committee

Chair
George Meyer gmeyer@celsiainc.com

Technical Chair
Ross Wilcoxon ross.wilcoxon@collins.com

Finance Chair
Jim Wilson jsw@raytheon.com

Steering/Technical Committee

Dereje Agonafer	agonafer@uta.edu
Joshua Gess	Joshua.Gess@oregonstate.edu
Genevieve Martin	genevieve.martin@signify.com
Bill Maltz	wmaltz@ecooling.com
Veerendra Mulay	vmulay@fb.com
Alfonso Ortega	alfonso.ortega@villanova.edu
John Parry	john_parry@mentor.com
Dave Saums	dsaums@dsa-thermal.com
Bernie Siegal	bsiegal@thermengr.net
Tom Tarter	ttarter@pkgscience.com
Winston Zhang	winstonzhang@novark.com.cn

SEMI-THERM Marketing
Denise Rael drael@semi-therm.org

SEMI-THERM Exhibits/Registration:
Bob Schuch rschuch@semi-therm.org

Graphic Design:
William Schuch bill@billschuch.net

SEMI-THERM 35 TOPIC CHAMPIONS AND PROGRAM/REVIEW COMMITTEE

Sharon Adam	Cisco	Bonnie Mack	Ciena
Sai Ankireddi	Maxim Integrated	Genevieve Martin	Signify
Janna Behm	ChargePoint	George Meyer	Celsia
Cathy Biber	Intel	David Nelson	Nelson Acoustics
Mark Carbone	Intel	Pritish Parida	IBM
Zeki Celik	Avago	James Petroski	Mentor
Marcelo del Valle	Intel	Adriana Rangel	Cisco Systems
Valerie Eveloy	Petroleum Institute, UAE	Peter Rodgers	Petroleum Institute, UAE
Angie Fan	Intuitive Surgical	David Saums	DS&A LLC
John Fernandes	Facebook	Mohammad Reza Shaeri	Advanced Cooling Technologies
Angel Han	Huawei Technologies	Jason Strader	Laird
Pablo Hidalgo	Aavid	Ross Wilcoxon	Collins Aerospace
Hussameddine Kabbani	Facebook	Jim Wilson	Raytheon
Taravat Khadivi	Qualcomm	Kazuaki Yazawa	MicrosanJ
Wendy Luiten	WLC Wendy Luiten Consultancy		

SEMI-THERM 35

www.semi-therm.org

Short Courses Monday, March 18, 2019

Short Course 1   Morning   8:00 a.m. – 12:00 p.m.
Statistical Analysis Methods for Dealing with Uncertainty in Thermal Testing
Ross Wilcoxon, PhD, Principal Mechanical Engineer, Mission Systems, Collins Aerospace

Statistical analysis is a methodology for using probabilistic methods to address the uncertainty that is inherent to all data. This course will give an overview of fundamental statistical methods that are used to identify the useful signals within a data set that may otherwise be obscured by the noise of data uncertainty. The class will provide the attendees with a better understanding of the how and why various statistical approaches are used as well as give tutorials on how to use a number of statistical analysis methods on actual data. About the Instructor

Ross Wilcoxon is a Principal Mechanical Engineer in the Rockwell Collins Advanced Technology Center. He conducts research and supports product development related to component reliability, electronics packaging and thermal management of avionics. Prior to joining Rockwell Collins in 1998, he was an assistant professor at South Dakota State University.

Short Course 2   Morning   8:00 a.m. – 12:00 p.m.
Design and Optimization of Heat Sinks
Dr. Georgios Karamanis, Co-Founder and Senior Engineer, Transport Phenomena Technologies, LLC

This course provides the audience with an understanding of heat sink design and optimization in the context of the thermal management of electronics. The course has two parts. The first part begins with an overview of common methods to manufacture heat sinks such as extrusion, die casting and forging, and discusses their advantages and disadvantages with respect to cost and fin geometry. Attention then shifts to the theory of spreading resistance and how it can be calculated in order to properly size the thicknesses of the bases of heat sinks. Next, the theory of the operation of heat pipes in tubular and flat (vapor chamber) configurations is presented along with their roles in smoothing out temperature gradients in the fins and bases of heat sinks.

In the second part of the course, single-phase conjugate heat transfer, where conduction in the heat sink is coupled to convection in the coolant, i.e., air or water, flowing through the heat sink is highlighted. We discuss why the constant heat transfer coefficient assumption tends to be an invalid one in real heat sinks by using specific examples. Then, the use of computational fluid dynamics (CFD) to compute conjugate Nusselt numbers is considered.

Next, we discuss how to embed pre-computed CFD results for conjugate Nusselt numbers and dimensionless flow resistances for heat sinks in flow network models (FNMs) of circuit packs such as blade servers.

Finally, a case study is presented where the fin height, length, spacing and thickness for 6 longitudinal-fin heat sinks cooling 6 microprocessors are simultaneously optimized by embedding the FNM representation of the circuit pack in a multi-variable optimization scheme.

Dr. Georgios Karamanis is a Co-Founder and Senior Engineer at Transport Phenomena Technologies, LLC. He received his Ph.D. and M.S. in Mechanical Engineering from Tufts University. He has expertise in analytical, numerical and experimental technics relevant to convective transport. He is the PI in a NSF Phase I SBIR awarded to Transport Phenomena Technologies, LLC, to develop specialized thermal modeling software for Data/Telco centers.

www.semi-therm.org

SEMI-THERM 35

Short Courses Monday, March 18, 2019

Short Course 3   Morning   8:00 a.m. – 12:00 p.m.
A Holistic Approach to Improve Mission Critical Facility Performance
Kourosh Nemati, Application Engineer, Future Facilities Ltd.

In recent years, data center designers & operators have focused on energy consumption, particularly PUE, to decrease operating expenses (OPEX). Hybrid cooling solutions, containment, and air-side or water-side economizers are examples of solutions implemented in data centers to achieve lower PUE. While these solutions have a positive effect on OPEX, they can also increase Capital Expenses (CAPEX) significantly. Meanwhile, a major driver of increased OPEX and PUE continues to go largely unnoticed – the fact that IT equipment, cooling infrastructure and data halls are all designed and tested separately. Since all these processes operate independently, it is a tremendous challenge to integrate them into one tool. However, if this can be achieved, data center energy consumption can be improved while providing sufficient cooling for IT, even during critical failure.

This entry-to-intermediate-level short course will demonstrate a comprehensive "Chip to Facility" CFD modeling process, using the Future Facilities software platform. The course will cover the entire process of detailed server modeling and room-level modeling, including different types of cooling strategies and control systems both in design and operational planning. Additional topics will be presented, including: a standard for data center model calibration, model integration to DCIM/ITSM via API web services, and an overview of external (generator yard and rooftop) modeling.

Kourosh Nemati is an application engineer at Future Facilities. He received his doctoral degree from the State University of New York at Binghamton. During his Ph.D., he has been involved in several data center thermal management projects, specialized in transport in data centers using both empirical and numerical approaches from server to room levels. He is a member of ASHRAE TC9.9, Green Grid and the NSF ES2 research project.

Short Course 4   Morning   8:00 a.m. – 12:00 p.m.
Thermal Challenges in Automotive Electronics
Tobias Best, Managing Director, Alpha-Numerics GmbH Course Description

Growing demands on electronic equipment in the automotive industry means a very precise consideration of thermal management is required. For several decades, there has been a trend that the performance increases, but the equipment gets smaller, leading to higher packing density. In addition to this challenge, which is common in other electronics industry segments, the automotive industry offers yet another hurdle. The installation space for the electronic equipment is usually not a simple boundary condition from a thermal point of view.

The thermal impact from solar radiation, noise-insulation (which acts as heat-insulation) and the effect of engine heat on electronics installed in the engine compartment all need to be considered. Without considering these effects in the design of the thermal management, the equipment might work as a prototype, but could completely fail in the field.

This short course will give an overview of the challenges an engineer will face when developing electronic equipment especially for the automotive industry. The course will concentrate on the physical area around thermal management and will show examples of the many challenges faced. The use of simulation to visualize the thermal behavior of the design and the creation of a digital twin for virtual prototyping will also be covered. The seminar will highlight some physical background concerning electronics cooling and will give ideas to help meet the latest requirements.

With more than 20 years experience using an industry specific CFD simulation tool and working as a consultant for the automotive industry, **Tobias Best** is currently owner and Managing Director of Alpha-Numerics GmbH in Germany.

www.semi-therm.org

SEMI-THERM 35

www.semi-therm.org

Short Courses Monday, March 18, 2019

Short Course 5   Afternoon   1:30 p.m. – 5:30 p.m.
Introduction to the Design and Implementation of Indirect Liquid Cooling for Electronic Systems
Alfonso (Al) Ortega, Ph.D., Professor and Director, Laboratory for Advanced Thermal and Fluid Systems, Villanova University
Rahima Mohammed, Senior Principal Engineer, Intel Corporation

The capacity of liquid cooling systems to manage heat dissipation from electronics far exceeds the capacity of air-cooled systems, a fact that has been known and pursued for decades. The preference for air cooling is readily justified because of ease of use and compatibility with electronics and their reliability. Air-cooling performance is ultimately limited by volumetric constraints on the size of the extended surface heat sink attached to high power components, acoustic limits on the allowable volumetric flow rates, and availability of air-movers that can deliver flow at pressure heads sufficiently high to overcome the pressure drop in volumetrically dense finned structures. Practically speaking, air cooling strategies cannot achieve heat sink resistances much below 0.1 C/W and component heat dissipations much greater than 100 W. Transitioning to liquids such as water or refrigerants as the primary heat transfer medium requires more exacting design and adaptation of infrastructure at system and component levels to accommodate delivery of liquid flow to high power devices.

This short course is intended for engineers who want to better understand strategic considerations in the selection of indirect liquid cooling solutions as compared to air-cooled solutions. The course will focus on the design and performance considerations for indirect (cold-plate based) liquid cooling solutions that use either single phase (liquid) or two-phase (boiling) convection as the primary strategy for heat removal. Topics to be covered include the following:
• Design drivers for liquid cooling transition in different platforms: Server, Desktop, Mobile
• System ramifications and trade-offs of solutions using liquid versus air cooling
• Design principles for single phase liquid-cooled cold plate design at conventional scales
and emerging principles and data for micro-scale heat sink design
• Understanding the behavior of boilers/evaporators with mini or microscale features
• Design principles for liquid cooling systems and their implementation

Dr. Alfonso Ortega is the James R. Birle Professor of Energy Technology at Villanova University. He is the Director of the Laboratory for Advanced Thermal and Fluid Systems and the Founding Director of the Villanova site of the NSF Center for Energy Smart Electronic Systems (ES2) founded in 2011. He received his B.S. from The University of Texas-El Paso, and his M.S. and Ph.D. from Stanford University, all in Mechanical Engineering. He was on the faculty of the Department of Aerospace and Mechanical Engineering at The University of Arizona in Tucson for 18 years. For two years, he served as the Program Director for Thermal Transport and Thermal Processing in the Chemical and Transport Systems Division of The National Science Foundation, where he managed the NSF's primary program funding heat transfer and thermal technology research in U.S. universities.

Dr. Ortega is a teacher of thermal sciences and experimental methods. He is an internationally recognized expert in the areas of thermal management in electronic systems. He has supervised over 40 M.S. and Ph.D. candidates to degree completion, 5 postdoctoral researchers, and more than 70 undergraduate research students. He is the author of over 300 journal and symposia papers, book chapters, and monographs and is a frequent short course lecturer on thermal management and experimental measurements.

He is a Fellow of the ASME and received the 2003 SEMITHERM Thermie Award and the 2017 ITHERM Achievement Award in recognition of his contributions to the field of electronics thermal measurements.

Rahima Mohammed is a Senior Principal Engineer and serves as the lead of the Customer Delight Office for strategic customers in Performance, Power and Competitive Analysis (P2CA) team of Intel Corporation. She has been with Intel over 20 years after graduate schooling from Yale. Before joining P2CA, she served as the Data Center customer solutions technologist and led data mining efforts on customer returned parts and as test and validation lead for server products in Manufacturing Validation Engineering (MVE). She also served as the advanced test module technologist in Manufacturing Development Organization (MDO). Prior to that, she served as the path finding czar for strategic emerging technologies across market segments and also setup the innovation programs for the division. Rahima led the team to deliver 15 advanced validation platform designs and pioneered innovative temperature margining thermal tools for over thirty-five silicon products. She also chairs various technical steering committees and serves on Industry advisory boards. She demonstrates consistent leadership in IP creation, and has published 100+ papers in Intel internal and external conferences and filed 5 patents. She serves as a reviewer for various conferences like Itherm, Interpack, and a program committee member of IEEE Semi-therm conference and Burn-in-test strategies workshop. She served as the vice-program chair, program chair, and general chair of Semi-therm conferences in 2014, 2015, and 2016, respectively. She has served as the senior advisor for Women at Intel Network of Guadalajara, Mexico for the past 8 years. She has been working with GHC and AnitaB since 2011.

www.semi-therm.org

SEMI-THERM 35

Short Courses Monday, March 18, 2019

Short Course 6   Afternoon   1:30 p.m. – 5:30 p.m.
Design of Experiments (DOE) for Thermal Engineering
James Petroski, Principal Consultant, Design by Analysis Technical Consulting

This course is intended to introduce people to the concept of Design of Experiments (DOE) and how it can be applied to engineering for effective design and experimentation. Beginning with a discussion of effective experimentation, the class will progress through different types of experimentation used today, the role of statistics in planning experiments and the product designs they influence, to an overview of various types of DOE's.
In depth presentation of certain DOE types will be given and the reason why the DOE type is chosen for a particular situation. The course will then show the process of setting up a "typical" DOE and follow with two examples, one from an analytical design using a DOE and a second of an experimental DOE of a system.

James Petroski is the founder and Principal Consultant of Design by Analysis Technical Consulting. Mr. Petroski has been involved in thermal, shock and vibration management of electronics systems for DOD, NASA and commercial applications with over 35 years' experience in the field of electronics packaging and LED thermal management. He received his Bachelors in Engineering Science and Mechanics from Georgia Tech and a MS degree in Engineering Mechanics from Cleveland State University. He has authored numerous papers related to LED and electronics packaging, has over thirty patents pertaining to solid-state lighting and electronics cooling, and is currently a member of the ASME K-16 Subcommittee on Heat Transfer in Electronics.

SEMI-THERM takes place at:
DoubleTree by Hilton San Jose
2050 Gateway Place, San Jose, CA 95110
Phone: 1 (408) 453-4000

For program details, registration, exhibition and hotel information visit
WWW.SEMI-THERM.ORG today!

---

www.semi-therm.org

Keynote Speaker
Tuesday March 19, 2019 9:10a.m. – 10:10a.m.

Challenges in the CPU and GPU Markets

The technical challenges for CPU and GPU products have gone through several inflections over the last 25 years. This talk will focus on these inflections as well as look forward to what may lie ahead.

Tom Dolbear, AMD

Since October 2017, Tom has been the Senior Director in AMD's Radeon Technology Group leading the global board hardware engineering team responsible for electrical design, power regulation, and thermal/mechanical design for GPUs and graphics cards for the mobile, gaming, and datacenter/machine learning markets. From 2009 to October 2017, Tom directed AMD's global packaging organization, developing solutions for PlaystationTM 4 and XboxTM One, for Fiji and Vega GPUs utilizing 2.5D technology, and for the EpycTM and RyzenTM processors. Tom was the catalyst behind the integration of the AMD and ATI packaging teams in 2008. From 1995 to 2009, he worked in several architecture, packaging, and platform engineering roles within AMD, including being a key contributor to bringing OpteronTM to the server market, the achievement of the first 1GHz CPU, and the development of AMD's first unique motherboard infrastructure for K7 CPUs.

Prior to joining AMD in 1995, he was a Member of Technical Staff at MCC, the first pre-competitive research consortium in the United States. He holds fifteen patents in the field of electronic packaging. He graduated from The University of Texas at Austin with a BS in Mechanical Engineering and from Stanford University with a MS in Mechanical Engineering.

Luncheon Speaker
Tuesday March 19, 2019

The Future of Innovation – Fusing Art and Technology

Domhnaill Hernon

Innovation is one of the most overused buzz words in modern society. If everything is "innovative" then surely nothing is?! Another popular mantra nowadays is to bring the humanities into the tech world for increased revenue owing to diverse perspectives. However, these are examples of popular sound bites that generate "check-the-box" exercises and thus limit our ability to progress humanity through technological evolution. In this talk I discuss the need to deeply understand what innovation truly is (and is not) and I share real examples of ways to develop innovative solutions by fusing art and technology.

Domhnaill Hernon is Head of Experiments in Arts and Technology (E.A.T.) at Nokia Bell Labs. He graduated with a B.Eng in Aeronautical Engineering, a Ph.D in fundamental fluid mechanics from the University of Limerick and an Executive M.B.A. from Dublin City University, Ireland. He is passionate about turning research/ideas into reality and exploring the bounds of creativity to push the limits of technology. Domhnaill was previously responsible for turning Bell Labs disruptive research assets into proto-solutions that could be tested at scale in the market and he established new methods to overcome the "Innovation Valley of Death". He is currently responsible for Bell Labs global activities in E.A.T. where he collaborates with the artistic/creative community to push the limits of technology to solve the greatest human need challenges.

Luncheon Speaker
Wednesday March 20, 2019

The Origins of Silicon Valley: Why and How It Happened Here

Presenter: Paul Wesling
IEEE Life Fellow

"Someday you will see Palo Alto blooming with nearly all the flowers of the earth and the fruit and shade trees of every zone... In the future we shall can this fruit and send it all over the globe in exchange for wealth..."

-Leland Stanford

Why did Silicon Valley come into being? The story goes back to local Hams (amateur radio operators) trying to break RCA's tube patents, "angel" investors, the sinking of the Titanic, Fred Terman and Stanford University, local invention of high-power tubes, WW II and radar, William Shockley's mother living in Palo Alto, and the SF Bay Area infrastructure that developed – these factors pretty much determined that the semiconductor and IC industries would be located in the Santa Clara Valley, and that the Valley would remain the world's innovation center as new technologies emerged – computers, then software, mobile, biotech, Big Data, VR, and now autonomous vehicles – and it would become the model for innovation worldwide.

Paul Wesling, an IEEE Life Fellow and Distinguished Lecturer, has observed the Valley for decades as an engineer, executive, resident, and educator, and has presented this talk world-wide. In this non-technical presentation, he gives an exciting and colorful history of device technology development and innovation that began in Palo Alto, then spread across the Santa Clara Valley during and following World War II. You'll meet some of the colorful characters – Leonard Fuller, Lee de Forest, Bill Eitel, Charles Litton, Fred Terman, David Packard, Bill Hewlett, Russ Varian and others – who came to define the worldwide electronics industries through their inventions and process development. You'll understand some of the novel management approaches that have become the hallmark of tech startups and high-tech firms, and the kinds of engineers/developers who thrive in this work environment. You'll handle an original Audion tube, invented by Lee de Forest and improved by him in Palo Alto. Paul will end by telling us about some current local organizations that keep alive the spirit of the Hams, the Homebrew Computer Club, and the other entrepreneurial groups where geeks gather to invent the future.

Embedded Tutorial
Wednesday March 20, 2019

MODELING TWO-PHASE HEAT TRANSFER SYSTEMS, PUMPED AND PASSIVE DESIGNS

Presented by:

George Meyer	Pritish R. Parida	Sobo Sun
Celsia	IBM Thomas J Watson Research Center	Celsia

Two-phase heat transfer systems utilize the latent heat property of a coolant fluid to transfer heat loads to regions or components where it could be efficiently dissipated to the ambient environment. The development of two-phase cooling for both two-dimensional (2D) and three-dimensional (3D) integrated circuits using pumped dielectric coolant and passive designs such as heat pipes, vapor chambers, thermosyphons, etc., has gained recent attention due to the ability to manage high heat densities, compatibility with electronics and above ambient temperature operation to achieve very low cooling energy usage. Development of this approach requires high fidelity and computationally manageable conjugate thermal models both at the device level as well as at the system level. This talk will describe a few modeling methodologies demonstrating the process of design and development of both passive and pumped two-phase heat transfer systems.

George Meyer is a thermal industry veteran with over three decades of experience in electronics thermal management. He currently serves as the CEO of Celsia Inc., a design and manufacturing company specializing in custom heat sink assemblies using heat pipes and vapor chambers. Previously, Mr. Meyer spent twenty-eight years with Thermacore in various executive roles including Chairman of the company's Taiwan operations. He holds over 70 patents in heat sink and heat pipe technologies and serves as a chairperson for the SEMI-THERM thermal conferences in the San Francisco area.

Pritish R. Parida received the B.Tech. degree in mechanical engineering from IIT Guwahati, Guwahati, India, in 2006, the M.Sc. degree in mechanical engineering from Louisiana State University and Agricultural and Mechanical College, Baton-Rouge, LA, USA, in 2007, and the Ph.D. degree in mechanical engineering from the Virginia Polytechnic Institute and State University (Virginia Tech), Blacksburg, VA, USA, in 2010. He is currently a Research Staff Member at IBM T. J. Watson Research Center, Yorktown Heights, NY, USA, where he develops new techniques and innovative solutions, offering fundamental breakthroughs in the state of the art to provide market differentiating technology for IBM's portfolio of products and services in the field of information technology. He addresses the thermal challenges in computer systems to achieve highly energy-efficient thermal designs to reduce the cooling energy used by computers in data centers. He has co-authored over 50 peer-reviewed publications and holds over 40 issued patents.

Sobo Sun is a 20 year thermal industry veteran with 80 patents in this field and expertise in heat sink modeling and design for manufacturability.
Prior roles include various senior management positions with Coolermaster and Thermacore.
Masters Mechanical Engineering, National Chung-Hsing University and PhD. ME candidate, National Chiao Tung University.

Evening Tutorial
Tuesday March 19 2019, 7:30 p.m.

Ai Is Helping Engineers Break Through The Barriers Of Thermal Design

Presenter:

Lieven Vervecken
Diabatix

Imagine what would happen if you explained the laws of thermodynamics to a hyperintelligent machine. With the capacity to think, imagine how that machine would optimally design cooling fins and cooling channels to precisely fit your needs. Would it decide to use parallel cooling fins, S-shaped cooling channels, or something new? Would it try to maximize the contact surface area, or not? Artificial intelligence is revolutionizing the way thermal engineers design cooling solutions. Discover how in a presentation by Diabatix.

Lieven Vervecken is CEO and co-founder of Diabatix, an engineering company specialized in advanced thermal design. Prior to founding Diabatix, Lieven received a PhD in mechanical engineering from the renowned University of Leuven, in the field of numerical simulations. Lieven incorporated his expertise into the advanced A.I. technology that lies at the heart of Diabatix. What started out as a small venture has become a fast-growing company serving multinationals all over the world.

Lieven is an experienced speaker at national and international conferences with a passion for the limitless possibilities of combining engineering with artificial intelligence technology. During his talk he will expound on the many doors the Diabatix technology can open.

Teardown Session
Tuesday Afternoon, March 19, 2019 4:45 p.m.

Alternative Thermal Solution For A Wireless Home Router

Presenter:
Justin Dixon, Electronic Cooling Solutions, Inc.

In recent years the demand for faster internet speeds has been steadily increasing. As consumers rely more on the internet for shopping, streaming videos, gaming, personal communications, business, etc. the technology used to support these applications must advance to keep in step with demand. The market dictates that devices of this nature be small, silent and aesthetically pleasing while also meeting performance requirements. Currently, many wireless devices use vents and heatsinks to cool internal components which alters the external appearance and can reduce aesthetic quality.

This Teardown session will explore the possibility of removing the heatsinks and vents currently used for cooling in a high end wireless home router and replacing them with an alternative thermal solution using heat spreaders and conductive enclosure materials. Thermal analysis and thermal test data are presented and used to demonstrate the functionality of the alternative thermal solution. A "tear down" and evaluation of the current thermal solution are presented. The evaluation includes IR scans to determine hot spots and thermal test data to determine component temperatures under stressed loading conditions. Illustrations of system components and architecture and obstacles to designing a thermal solution are also discussed.

Justin Dixon has been a consulting thermal engineer at Electronic Cooling Solutions for 5 years. He has provided thermal analysis and testing services in the consumer electronics, telecommunications, automotive and medical devices industries. Recently he has consulted in the design for a number of consumer electronics devices such as routers, access points, cameras, LCD displays and more. He graduated, with honors, from San Jose State University with a degree in Mechanical Engineering and a focus on thermal fluids.

THERMI Award

Each year, SEMI-THERM honors a person as a Significant Contributor to the field of semiconductor thermal management. The THERMI award is intended to recognize a recipient's history of contributions to crucial thermal issues affecting the performance of semiconductor devices and systems.

The voting body of past THERMI winners and the current year General Chair are pleased to present the 2019 THERMI Award to:

Dr. Peter Raad
Southern Methodist University

Dr. Raad will be giving a presentation:
'Reflections on a Journey of Developing Means to Characterize Hot Spots in Cool Chips'

Peter E. Raad is a professor of mechanical engineering at Southern Methodist University (SMU) in Dallas, Texas. He first joined SMU in 1986 and has previously served as the associate dean of its School of Engineering. From 2000 to 2012, he founded and directed the Hart eCenter at SMU, a university-wide center to address the impact of the interactive networked technologies on society. During that time, he also founded and directed The Guildhall at SMU, a first of its kind, graduate program in digital game development. Raad has received over $2.8 million in funding support for his research in tsunami mitigation and in metrology of submicron electronics. In 2006, he founded TMX Scientific, a company to innovate and commercialize deep submicron thermal measurement systems and ultrafast thermal computational engines. Raad's work in the thermal management field includes the development of innovative deep-submicron thermal metrology techniques and systems, as well as novel coupling of computations and measurements to provide transient, three-dimensional temperature fields in electronic structures with inaccessible internal features.

His honors include the Allan Kraus Thermal Management Medal (2014); the Harvey Rosten Award for Excellence in the Physical Design of Electronics (2006); the ASME North Texas Section Engineer of the Year (1999-2000); the Next-Gen's Top 25 People of 2007 (most influential in the video gaming industry); and Outstanding Graduate (four times) and Undergraduate (three times) Faculty Awards at SMU.

He has published over 55 journal articles and given more than 100 conference and invited talks. He holds U.S. and international patents in thermal metrology and computational characterization of multiscale integrated circuits. He is a Fellow of ASME and a Senior Member of IEEE. He received his BSME (with honors, 1980), MS (1981), and PhD (1986) in mechanical engineering from the University of Tennessee - Knoxville.

SEMI-THERM® 35

Thursday March 21, 2019 12:30 p.m.

The 2018 Harvey Rosten Award

Compact Cooling-System Model for Transient Data Center Simulations
Jim VanGilder, Chris Healey, Wei Tian, Michael Condor, Quentin Menusier
For Outstanding Work in the Field of Thermal Analysis of Electronic Equipment.

Jim VanGilder received a BS from the University of Maine (1992) and an MS from Duke University (1993) in Mechanical Engineering. He joined Flomerics in 1997 where he initially focused on CFD for traditional electronics thermal applications and later on building-scale applications and, ultimately, data centers. He joined Schneider Electric (APC at the time) in 2003 where he focused on developing practical and fast tools to assist the thermal design and operation of data centers; Jim currently directs CFD development and related research. He has authored over 50 technical publications and holds more than 30 US patents related to data-center and electronics cooling. Jim is a long-time member and former chair of ASHRAE TC 4.10, Indoor Environmental Modeling, and a frequent contributor to the ASME InterPack and IEEE ITherm conferences. Jim is also a licensed professional engineer in the state of Massachusetts.

Chris Healey is a graduate of the College of William and Mary (2005, BS Math) and Georgia Institute of Technology (2010, PhD Industrial Engineering). He is a Data Science Team Leader in Schneider Electric, working in thermal analytics and data science in the goal of efficiency and reliability of data center systems through optimal design, efficient control, and predictive maintenance. He has authored or co-authored seven journal papers and numerous conference proceedings.

Michael Condor received a BS from Bucknell University (2000) in Computer Science. Michael has nearly 20 years of experience developing software for data center management technology (DCIM) as well as thermal analytics software. Michael is currently the lead software engineer for the Thermal Analytics team in Schneider-Electric.

Wei Tian received a PhD in Civil Engineering from the University of Miami. He is currently a research engineer at Schneider Electric, working in developing numerical models and tools to analyze thermal performance of data centers. His research interests include computational fluid dynamics, energy system modeling, and dynamic simulation and optimization. Wei has authored over 20 technical papers.

Quentin Menusier received an MS from ENSMM (2018), a French Engineering School in Computational Mechanics. He had the opportunity to work in Schneider Electric as a Computational Fluid Dynamics intern. He currently works in France as an Engineering Consultant.

The Harvey Rosten Award
The Award is for outstanding work, recently published or in the public domain, which advances the analysis or modeling of thermal or thermomechanical effects in electronic equipment or components, including experiments aimed specifically at the validation of numerical models. The award is in the form of a plaque and a $1000 cash prize. The Award was established by the family and friends of Harvey Rosten, to commemorate his achievements in the field of thermal analysis of electronics equipment, and the thermal modeling of electronics parts and packages. The Award is made annually to encourage innovation and excellence in these and closely related fields.

The recipient is selected by the Selection Committee, made up of eminent practitioners in the electronics-thermal field.
The criteria for selection are:
- The work represents an advance in thermal analysis or thermal modeling of electronics equipment or components, including experiments aimed specifically at validating numerical models.
- The work demonstrates clear application to practical electronics design.
- The work demonstrates insight into the physical processes affecting the thermal behavior of electronics components, parts and systems.
- The work is innovative in embodying this understanding in either thermal analysis or thermal modeling.
- A pragmatic approach is taken in the application of the work.

Mechanical & Aerospace Engineering
The University of Texas at Arlington

ES2 — Center for Energy-Smart Electronic Systems

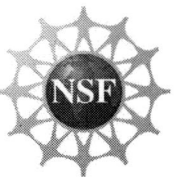 NSF

VILLANOVA
UNIVERSITY

celsia°
Making Hot Technology Cooler

We are proud to sponsor:

The SEMI-THERM Educational Foundation
Thermal Hall of Fame

Lifetime Achievement Award

Presented To

Márta Rencz
In Recognition of Significant Contributions
to the Field of Electronics Thermal Management

Márta Rencz received the Electrical Engineering degree, the Doctor in Engineering Degree and the PhD degree from the Budapest University of Technology and Economics, where she has also obtained the Habilitation. She has received the Doctor of Science degree from the Hungarian Academy of Science in the field of Microelectronics. She is a professor at the Budapest University of Technology and Economics. Between 2005 and 2013 she served as the Head of Department of Electron Devices. She has participated in numerous international research projects, mostly in the field of investigating, measuring and modeling multi-physical effects in electronics. She has published her theoretical and practical results in more than 300 technical papers.

She has been the guest editor of over 10 special issues of various scientific Journals in the fields of thermal investigations and thermal management in electronics. She is regularly reviewing scientific papers and international research proposals. She was a co-founder and CEO of Micred Ltd that is now part of Mentor, a Siemens business, where she still holds a research director position.

She initiated in 1992 the THERMINIC EU research project that has led to the THERMINIC workshops, dealing with thermal issues in electronics, giving a forum for thermal management experts. Today she is the chair of the steering committee of THERMINIC.

She holds various awards of excellence, among others Harvey Rosten award (2001) and the Allan Krauss thermal management award of ASME (2015). In 2013 she has received the Doctor Honoris Causa degree from the Tallinn University of Technology in Estonia.

TABLE OF CONTENTS
SEMICONDUCTOR THERMAL MEASUREMENT AND MANAGEMENT SYMPOSIUM

Welcome to SEMI-THERM 35 iii

SEMI-THERM Committees iv

SEMI-THERM Short Courses v

SEMI-THERM Keynote Speaker: Tom Dolbear, AMD ix

SEMI-THERM Luncheon Speakers: Domhnaill Hernon, Nokia; Paul Wesling, HP (retired) x

SEMI-THERM Embedded Tutorial: George Meyer, Sobu Sun, Celsia; Pritish Parida, IBM xii

SEMI-THERM Evening Tutorial: Lieven Vervecken, Diabatix xiii

SEMI-THERM Teardown: Justin Dixon, Electronic Cooling Solutions, Inc. xiv

Thermi Award: Dr. Peter Raad, Southern Methodist University xv

Rosten Award: Jim VanGilder, Chris Healey, Wei Tian, Michael Condor, Quentin Menusier xvi

Thermal Hall of Fame Award: Márta Rencz, Budapest University of Technology and Economics xvii

Session 1: LEDs
Chair: Jim Petroski, Mentor, A Siemens Business

A Methodology to Determine the Sites of Variability in an LED Assembly 1

Robin Bornoff[1], Thomas Mérelle[2], Josephine Sari[3]3, Alessandro Di Bucchianico[3], Gabor Farkas[1]
[1]Mentor, A Siemens Business, [2]Pi-Lighting, [3]Eindhoven University of Technology

Accurate Thermal Transient Measurements Interpretation of Monochromatic LEDs 7

Alexeev Anton[1], Genevieve Martin[2], Grigory Onushkin[2], Jean-Paul Linnartz[2]
[1]Eindhoven University of Technology, [2]Signify

Implementation of a Multi-domain LED Model and its Application for Optimized LED Luminaire Design 12

János Hegedüs[1], Gusztáv Hantos[1], Robin Bornoff[2], Márta Rencz[1,2], András Poppe[1,2]
[1]Budapest University of Technology and Economics, [2]Mentor, A Siemens Business

Session 2: Two Phase Cooling
Chair: George Meyer and Sobo Sun, Celsia Inc.

Assessment of Critical Heat Flux on Finite Size Surfaces Under Pool Boiling 18

Julia Reed, Vijay K. Dhir, University of California, Los Angeles

Molecular Dynamic Simulation of Evaporative Heat Transfer on Graphene Coated Silicon Substrate for Electronics Cooling 26

Binjian Ma[1], Rui Zhou1, Shan Li[1], Junhui Li[1], Damena Agonafer[1], Baris Dogruoz[2],
[1]Washington University in St. Louis, [2]Cisco Systems Inc.

Experimental and Numerical Investigation of Microdroplets Evaporation on Porous Pillar Structures *

Li Shan, Washington University in St. Louis

Heat Pumps to Upgrade Data Center Waste Heat: Integration with 2-Phase Cooling *

Steven Schon, QuantaCool

Thermal Performance of Metal Foam Heat Sink with Pin Fins for Non-Uniform Heat Flux Electronics Cooling 30

Yongtong Li[1,2], Liang Gong[1], Minghai Xu[1], Yogendra Joshi[2],
[1]China University of Petroleum, [2]Georgia Institute of Technology

* -- This presentation has no formal paper.

Session 3: Thermal Interface Materials
Chair: Jason Strader, Laird

Mechanical Cycling Reliability Testing of Thermal Interface Materials for Semiconductor Test 38
Dave Saums[1], Tim Jensen[2], Carol Gowans[2], Ron Hunadi[2], Mohamad Abo Ras[3],
[1]DS&A LLC, [2]Indium Corporation, [3]Berliner Nanotest und Design GmbH

Liquid Metal Innovations for High Performance TIMs *
Timothy Jensen, Indium Corporation

High Performance Lightweight Ceramic Material for Thermal Management in Electronic Devices 45
Bei Xiang, Chandra Raman and Xiang Liu, Momentive Performance Materials Quartz, Inc.

**Performance of Durable High-Performance Polymer Composite TIMs Under
Accelerated Aging Conditions** 48
Hyungyung Jo, John A. Howarter, Purdue University

Thermal Diffusivity Characterization of Thick Graphite Foils *
Rick Beyerle, Martin Smalc, Rajath Kantharaj, Jonathan Taylor, Julian Norley, NeoGraf Solutions

Session 4: Automotive/Aerospace/Outdoor
Chair: Hussamedine Kabbani, Facebook

**The Impact of Anodization on the Thermal Performance of Passively Cooled Electronic
Enclosures Made of Die-cast Aluminum** 51
Zhongchen Zhang[1], Michael Collins[2], Chris Botting[3], Eric Lau[3], Majid Bahrami[1],
[1]Simon Fraser University, [2]University of Waterloo, [3]Delta-Q Technologies

Development of a 3D Printed Loop Heat Pipe 58
Bradley Richard, William G. Anderson, Joel Crawmer, Advanced Cooling Technologies Inc

**Measurement of Thermal Resistance of Thermal Interface Materials with High In-plane Thermal
Conductivity using Transient Thermal Based Structure Function Analysis** *
Aloysius Davin Oetomo, Carbice Corporation

Session 5: CFD/Numerical Modeling
Chair: Taravat Khadivi, Qualcomm

**Design using Multi-Scale, Multi-Physics Analyses And Shape Optimization, for Compact
Heat Transfer Devices** *
Daniel Bacellar, Dennis Nasuta, Cara Martin, Reinhard Radermacher, Optimized Thermal Systems, Inc.

Research on Package Thermal Resistance of Power Semiconductor Devices 61
Koji Nishi, Ashikaga University

The Necessity for Thermal-Electrical-Multiphysics for Board Heating in a Server Rack Unit *
Jared Harvest, Wade Smith, Satyajeet Padhi, ANSYS, Inc.

**Temperature Profile of High Power Density (HPD) ASIC Device Mounted on Multi-layered
Diamond Enhanced Heat Spreader** *
Thomas Obeloer, Russell Mason, Daniel Twitchen and Firooz N. Faili, Element Six Technologies

Practical Evaluation of Thermally - Conductive Plastics and Guidelines for Use *
Dave Saums, DS&A LLC

Optimization of an Array of Heat Sinks to Satisfy an Arbitrary Objective Function *
Georgios Karamanis and Marc Hodes, Transport Phenomena Technologies, LLC

Session 6: Two Phase Cooling
Chair: Pritish Parida, IBM

CTE Matching Heat Pipe Thermal Ground Plane *
Nelson J. Gernert, Aavid Thermal Division of Boyd

* -- This presentation has no formal paper.

An Ultra-Thin Loop Heat Pipe with Long-Distance Heat Transport for Cooling of Small Electronic Devices 66

Shuto Tomita, Ai Ueno, Hosei Nagano, Nagoya University

Evaluation of the Performance of Various Heat Pipe Mounting Methods with Various Thickness TIM's and Mounting Pressures *

George A Meyer IV, Sobo Sun, Rock Chin, Celsia Inc

Relative Performance of Two-Phase vs Solid Conductive Heat Spreaders for High Heat Flux Applications 70

Joe Boswell, Corey Wilson, Josh Schorp, and Dan Pounds, ThermAvant Technologies

The Impact of Heat Rejection Architecture on the Thermal Performance of a Pumped Two-Phase Cooling System *

Elizabeth Baker, Danah Valez, Timothy A. Shedd, Florida Polytechnic University

Session 7: Consumer Electronics
Chair: Mark Carbone, Intel and Angel Han, Huawei

Analysis of Natural Frequency Dependency on Temperature Variation of MEMS Vibratory Gyroscope 76

Jacek Nazdrowicz, Andrzej Napieralski, Lodz University of Technology

Battery Discharge Capacity Calculation by Temperature Measurement 83

Jeevan Kanesalingam and Khoo Li Lian, Motorola Solutions

Exploring Heatpipe Configurations for Package On Package (PoP) Cooling 87

Sankarananda Basak[1], Ryota Watanabe[2], [1]Intel Corp, [2]Lenovo (Japan) Ltd.

Session 8: Data Center Cooling
Chair: Marcelo del Valle, Intel

Simulation-Based Optimization of Data Center Cooling Performance using Performance Indicators 91

John Petrongolo[1], Kourosh Nemati[2], and Kamran Fouladi[1], [1]Widener University, [2]Future Facilities

Transient Analysis Overshoot in Temperature for High Power Thermal Solutions 96

Javier Avalos and Enrique Barreto, Intel Corp.

Airflow Management using Active Air Dampers in Presence of a Dynamic Workload in Data Centers 101

Sadegh Khalili, Ghazal Mohsenian, Anuroop Desu, Kanad Ghose, Bahgat Sammakia, Binghamton University

---

SUBMIT A PAPER FOR SEMI-THERM 36!

As you further develop a technique or application, consider documenting it for the thermal community. **SEMI-THERM 36** will begin accepting abstracts during the summer (deadline is September 15, 2019). We welcome your submissions! Visit us at www.semi-therm.org.

SEMI-THERM 36 is March 16-20, 2020 – *be there!*

* -- This presentation has no formal paper.

A Methodology to Determine the Sites of Variability in an LED Assembly

Robin Bornoff[1], Thomas Mérelle[2], Josephine Sari[3], Alessandro Di Bucchianico[3], Gabor Farkas[4]

[1]Mentor Graphics, Mechanical Analysis Division, 81 Bridge Road, Hampton Court, Surrey UK
[2]Pi-Lighting, Avenue Ritz 19, 1950 Sion, Switzerland
[3]Eindhoven University of Technology, Department of Mathematics and Computer Science, P.O. Box 513, 5600 MB Eindhoven, The Netherlands
[4]Mentor Graphics, Mechanical Analysis Division, Infopark D, Gábor Dénes u. 2., Budapest, 1117, Hungary
[1]robin_bornoff@mentor.com

Abstract

Variations in the manufacturing assembly process together with any changes in materials sourcing can give rise to variation in the functional performance of an LED. These variations can lead to differences in thermal characteristics such as Zth (transient thermal impedance) and RthJC (thermal resistance between chip junction and case. This study aims to analyze the measured variations of a number of LED samples and explore a methodology that will provide insights into the root cause of those variations, In addition the statistical properties of the variations are applied in a Monte Carlo thermal modelling context. Such insights and modelling methods would be employed by a lighting engineer when selecting LEDs and when simulating their thermal dependent optical behavior.

Keywords

LED, Structure Function, Variation

Nomenclature

Rth — Thermal Resistance (K/W)
Cth — Thermal Capacitance (J/K)
Tj — Junction Temperature (K)
Ta — Ambient temperature (K)
P — Dissipated power (W)
Rja — Thermal Resistance between junction and ambient $= (Tj - Ta) / P$ (K/W)
Zth — Transient thermal response between junction and ambient $= (Tj - Ta) / P$ (K/W) as a function of time

1. Introduction

In response to the commoditization of LEDs, a European Union consortium of 15 partners, Delphi4LED, is seeking to identify and elaborate a set of methodologies and standards to better support the design of cost effective and reliable LED lighting solutions. This is done through a combination of measurement and simulation being used to extract thermal-optical-electrical multi-domain LED compact models. A modular approach in the application of these models [1], [2], [3], [4] would then be used by lighting designers to meet optical and reliability design goals.

The predictive accuracy of a compact model will be affected by any variability in the thermal performance of an LED part sample. Analyzing the variations in the measured thermal responses of multiple samples will provide insight into the physical location of the causes of the variations and thus enable more informed assumptions as to the root cause.

Measurement of a structure function, using the JESD 51-1 electrical test method and processing of the recorded temperature response, is well established. For LED devices, Mentor tools T3Ster [5] and TeraLED [6] are used to perform such measurements, conforming to the latest JEDEC LED testing standards and CIE recommendations; CIE 225:2017, JESD51-52 and JESD51-51 [9], [10], [11]. These standardised test methods involve measurement of the total emitted optical power of the LED which is then used to determine the thermally dissipated power within the LED. It is this thermal power that is then used to convert the measured thermal response into a Zth profile. As the operating (junction) temperature is a key factor in the behavior of any derived multi-domain compact model, these tests are conducted at known and maintained constant values.

In this study 11 samples of an LED part, a Royal Blue part held at a Tj of 75 degC and driven by a forward current of 700mA, are measured using the above methods. Observed differences in the resulting structure functions are related to differences in the package construction. Plotting a structure function in terms of a 'local' Rth vs. cumulative Cth (as opposed to the more classic cumulative Cth vs. cumulative Rth) is a novel approach that more readily indicates the distribution of Rja along the heat flow path. Increased levels of local variation in Rth can be correlated to offending part materials.

This study presents 2 methods that relate a location of increased variation on the heat flow path to the corresponding offending object; analytically by noting the cumulative Cth of objects from a calibrated 3D thermal model of the test assembly and a transient dual interface (TDIM) [12] approach that uses perturbed simulations. Having correlated the increased variations in local Rth with physical locations, underlying causes may be postulated [13].

Based on the statistical correlation of the variations of Rth in a 12 stage RC ladder model of the measured LEDs, a Monte Carlo modelling approach is applied to generate 1000 Zth responses, from which a quantified variation of Rja is determined.

2. Structure Function Variation Comparison

All 11 cumulative structure functions are shown in Figure 1.

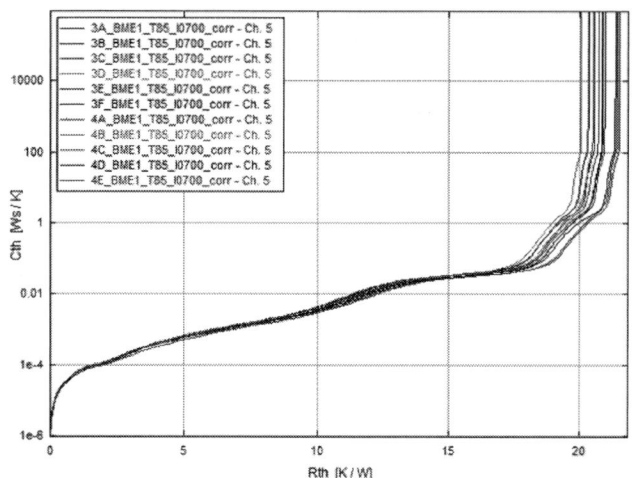

Figure 1: Cumulative Structure Functions of the 11 Measured LEDs

Rja standard deviation is +/- 2.1% around an average value of 20.87 K/W and a median value of 20.85 K/W. Any variation in thermal capacitance would manifest itself as scaling of the plots on the Cth (vertical) axis. This is not the case. This can be confirmed by plotting the local *Cth* values along each structure function (the difference between consecutive cumulative *Cth* points in Figure 1). These, together their normalized standard deviation, are shown in Figure 2.

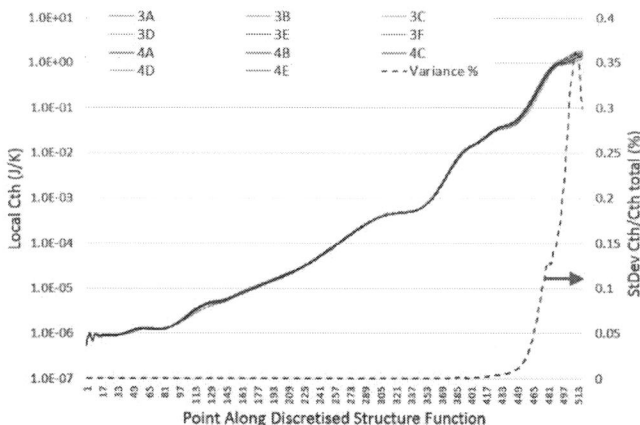

Figure 2: Local Cth Comparison of the 11 Measured LED Samples

Very low levels of variation in local Cth are seen, a maximum value of 0.35% observed right at the end of the heat flow path. It is evident that the variations between the 11 structure functions occur due to variation in the Rth along the heat flow path. This is seen by the structure functions scaling in the Rth (horizontal) axis in Figure 1. To better indicate these variations a novel Local *Rth* vs. (Average) Cumulative *Cth* plot is shown together with the corresponding normalized standard deviation in Figure 3.

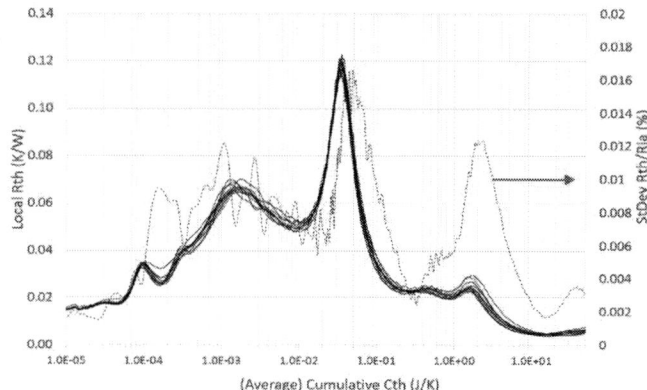

Figure 3: Local Rth vs. Cumulative Cth for 11 Measured LED Samples

Whereas the cumulative structure function plot of Figure 1 represents thermally resistive sections as longer horizontal sections of the structure function, the variation in local *Rth* shown in Figure 3 more clearly manifests as peaks. Corresponding peaks in the variation of local *Rth* indicate the locations that give rise to the overall variation of *Rja*. Now that the peaks in the variation of local *Rth* have been identified, the next step is to correlate those locations with the offending physical objects.

3. Analytical Mapping of Objects to Structure Function Features

A 3D thermal model of the entire tested assembly (LED, MCPCB and Coldplate) is constructed. FloTHERM [7] is used to construct the model, which is calibrated against the median LED sample [14]. The geometry of the FloTHERM model is shown in Figure 4.

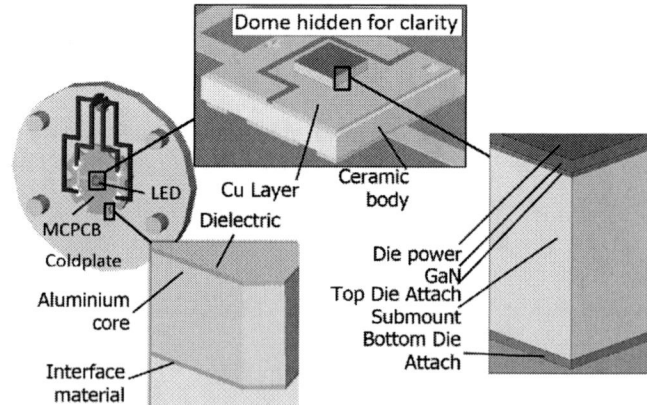

Figure 4: FloTHERM Model LED, MCPCB and Coldplate

Capitalizing on the predominant heat flow path from source, through the chip stake, package ceramic body, MPCB and Coldplate, the thermal capacitance along that path may be analytically calculated by considering the density, volume and specific heat of each of these pre-calibrated model objects. The *Cth* axis of the local *Rth* plot of Figure 2 can then be annotated with the corresponding objects (Figure 5).

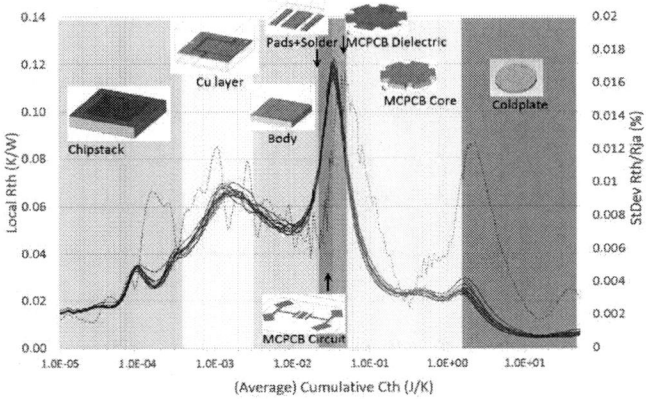

Figure 5: Local *Rth* vs. Cumulative *Cth* Object Annotated Comparison

Using this analytical approach, regions of increased variation are observed either side of the interface between the chip stack and the Cu layer, largest variations in the MCPCB dielectric layer and a distinct increase in variation at the interface between MCPCB and Coldplate.

4. Pertubation Mapping of Objects to Structure Function Features

JESD 51-14 [8], is a standardised method to determine Junction to Case thermal resistance (*Rjc*) for an IC package that exhibits a single predominant heat flow path. It involves 2 measurements where the material condition at the case face interface is changed. Comparison of the 2 measured structure functions identifies the *Rth* point at which they diverge from each other. This point is then taken to be *Rjc*.

A similar approach is taken in this study. The calibrated FloTHERM model is perturbed, not via the introduction of a resistive layer, but by (slightly) modifying the thermal conductivity of an object. Comparison of the local *Rth* structure functions before and after the conductivity change is then presented. The point at which they deviate indicates the 'start' location of that object (with respect to the heat flow path) on the structure function. In addition the '*Rth* span' of that object on the structure function is determined

This perturbation approach was applied at the 4 locations of increased local *Rth* variance seen in Figure 5 and noted in Section 2.

The thermal conductivity of the bottom die attach layer, between the submount and the Cu layer, was decreased by a small amount (so as to achieve a 2% increase in *Rja*) and the structure function comparison shown in Figure 6.

The (*Cth*) start point at of the perturbed object is taken to be the point at which the difference in local *Rth* exceeds 10% of the subsequent maximum change in local *Rth*. In this case *Cth* = 1.49e-4 K/W. It is noted that the variation in local *Rth* continues well beyond the assumed end point of the bottom die attach layer, to well within the Cu layer annotated band seen in Figure 5.

The next perturbation was the Cu layer on top of the LED package body ceramic. Local *Rth* perturbed structure functions is seen in Figure 7.

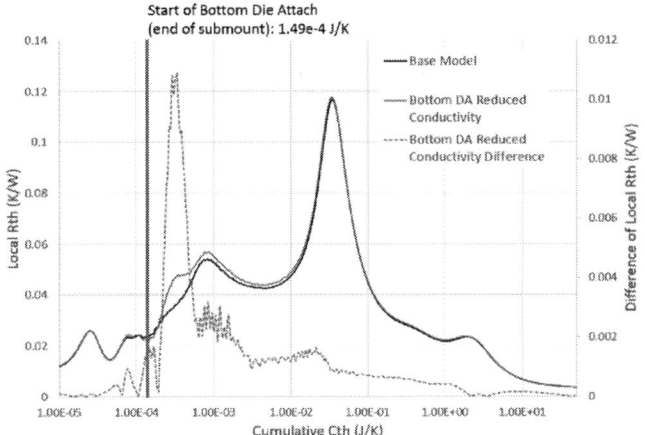

Figure 6: Perturbed Identification of Bottom Die Attach

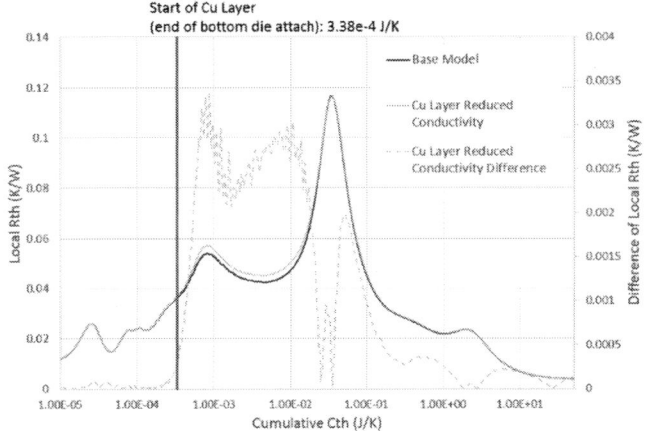

Figure 7: Perturbed Identification of Cu Layer Start

The start point of the Cu layer was at *Cth* = 3.38e-4 J/K. Again, the influence of the Cu layer is seen to extend well beyond where the object is expected to span from an analytical perspective. This agrees with the interdependency in fluctuation of local *Rth* as noted in [15].

Next the MCPCB dielectric layer was perturbed (Figure 8). The start point of the MCPCB dielectric layer was at *Cth* = 0.0172 K/W.

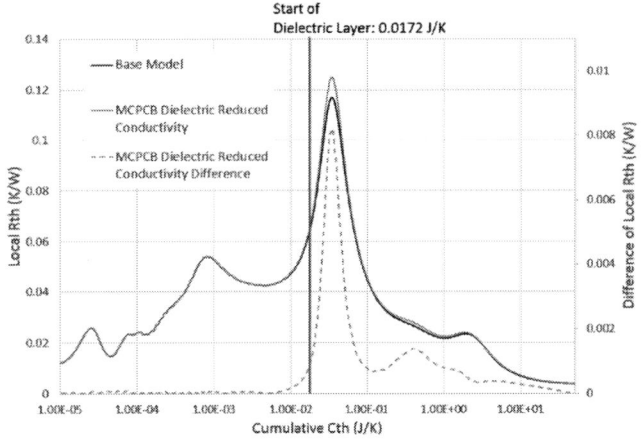

Figure 8: Perturbed Identification of MCPCB Dielectric Start

Finally, the interface between MCPCB and Coldplate was perturbed and the structure function comparison shown in Figure 9.

Although the start of the Coldplate was seen at 0.41 J/K (taking the 10% of maximum local Rth variation criteria), the effect of this interface was seen to start as 'early' as 0.02 J/K.

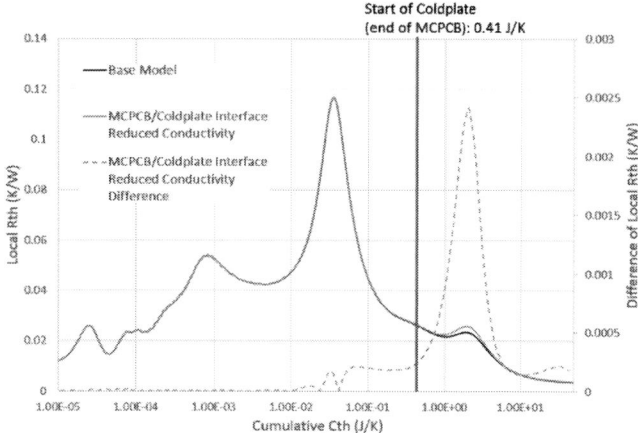

Figure 9: Perturbed Identification of Coldplate Start

5. Monte Carlo Modelling

In addition to providing insight into the root causes of the observed variation in thermal behavior, the 11 measurements can be used to derive a predictive model that is capable of also replicating the observed variation thermal behavior. A Cauer RC ladder type compact model topology is selected as this is readily extractable from the measured structure functions. A 12 stage ladder is considered. The stages of the ladder are artificially selected so that each stage spans a Cth region of similar measured variation.

A correlation coefficient matrix is calculated indicating which stage *Rth* value of which sample is correlated to what stage *Rth* values of the other samples. Value less than 0.75 are considered as uncorrelated (Figure 10).

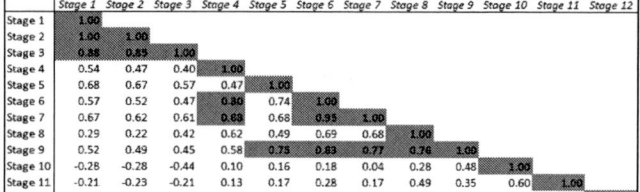

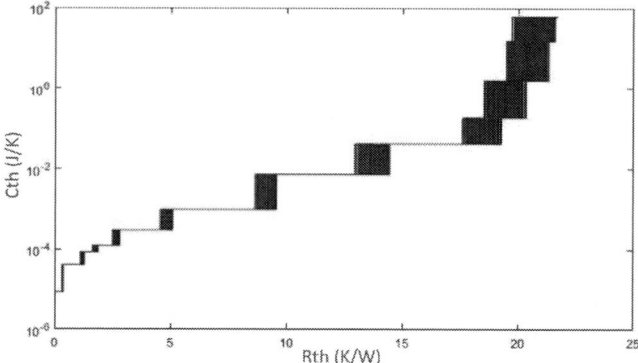

Figure 10: 11 Sample, 12 Stage Rth Values and Correlation Coefficient Matrix

A Monte Carlo simulation approach is taken to construct 1000 12 stage RC ladder models. The *Rth* of each stage is varied assuming a normal distribution of *Rth*. This is done independently for stages that show small to no correlation with other stages, i.e. stage 5, 8, 10, 11 and 12. For the other stages the *Rth* values are derived from the values of the stages they are correlated with (this is elaborated in [16]). The resulting 1000 12 stage RC models are shown in Figure 11.

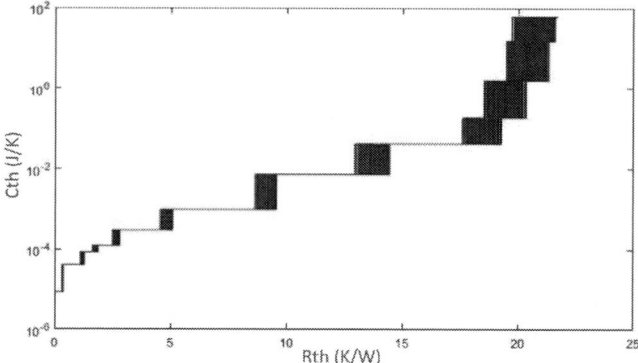

Figure 11: 1000 Monte Carlo 12 Stage RC Models (Median model shown in red)

1000 transient thermal response simulations are conducted and the resulting histogram distribution of *Rja* is shown in Figure 12.

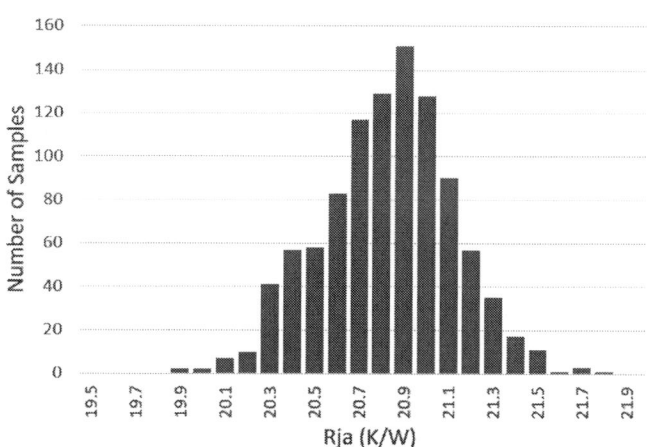

Figure 12: Histogram of *Rja* Distribution from 1000 Monte Carlo Simulations

6. Discussion

6.1. Variability Insights

Two methods are presented that seek to correlate the start point of an object on a local *Rth* structure function plot to the object itself. How closely these object start points match, as predicted by the 2 methods, is seen in Figure 13.

Generally the perturbation method predicts an object start point at a lower *Cth* value than the analytical approach. This is ascribed to the effects of heat spreading. The analytical approach is underpinned by the assumption that the temperature at each interface, throughout the transient thermal response, is uniform. Due to the effects of heat spreading this is not the case. The perturbation approach does not suffer this assumption and will therefore identify a deviation earlier in the thermal response, as soon as the heat starts to penetrate that object. The perturbation approach, although more computationally expensive, is considered more reliable.

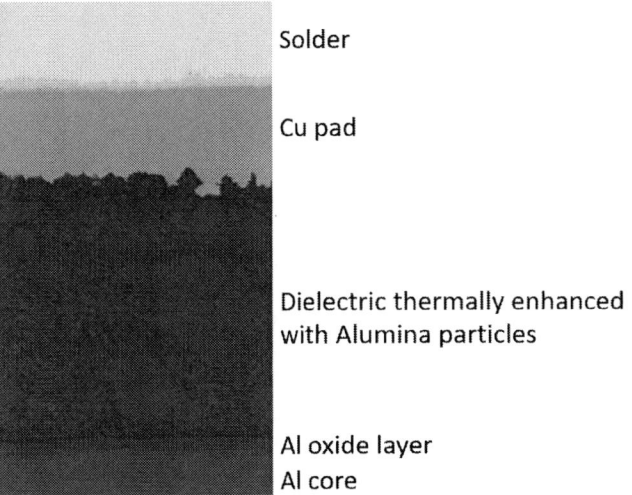

Figure 13: Comparison between Analytical and Perturbation methods (1: Submount/Bottom die attach interface, 2: Bottom die attach/Cu layer interface, 3: Circuit/MCPCB Dielectric interface, 4: MCPCB/Coldplate interface)

The distribution in local *Rth* variation along the heat flow path shown by Figure 3 can be considered as a super-position of the individual influences on variation shown in Figures 6, 7, 8 and 9.

The analytical and perturbation methods can be used to identify which object may be responsible for the observed levels of variation in local *Rth* but not the *reason* why. Some possible root causes are proposed based on knowledge of at least what objects are responsible.

An SEM image showing the bottom die attach layer is shown in Figure 14. This layer is seen to be voided. The amount of voiding is non-uniform, will be process dependent and therefore a likely candidate to explain the increased variation at this location (Figure 13 – point 1).

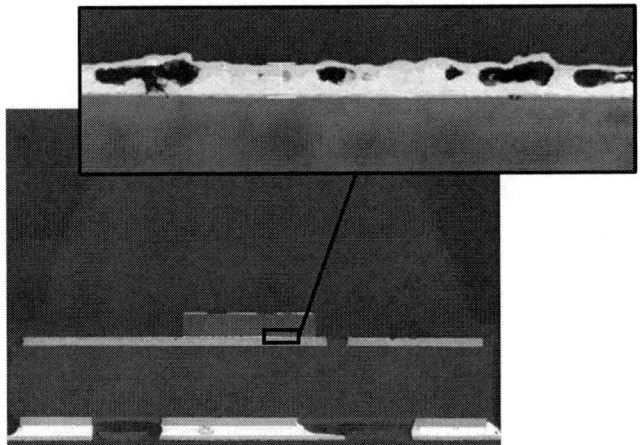

Figure 14: Voided Bottom Die Attach

The increased variation seen throughout the Cu layer, extending to the ceramic package body (Figure 13 – point 2 to point 3), is not readily explainable. As the thicknesses of those objects are well characterized, it must be assumed that

some variation in material thermal conductivity, surface oxidization or the existence of an unmodeled attach layer is to blame.

The largest variation in local *Rth* is seen in the object that is also contributes most to the overall *Rja* of the system, the MCPCB dielectric (Figure 13 – point 3). Such dielectric layers are often thermally enhanced via the addition of Alumina particles (typical section image shown in Figure 15). Variation in the distribution and volume fraction of these particles throughout the layer may well account for the observed *Rth* variation.

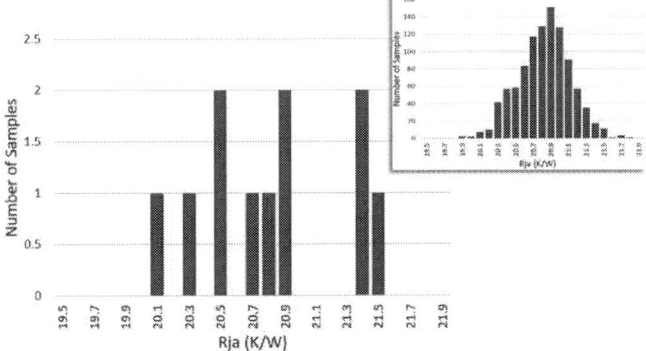

Figure 15: Typical Thermally Enhanced MCPCB Section

Finally, the increased variation at the MCPCB/Coldplate interface (Figure 13 - point 4) is attributed to the differences in the pressure exerted as the MCPCB was screwed down onto the Coldplate as this pressure was not guaranteed constant by the testing standards.

6.2. Monte Carlo Modeling

Determining dependable thermal characteristic statistical information from just 11 measured samples is not viable. The Monte Carlo approach, based on models derived from a small set of measurements, allows for a much wider (simulated) sample population to be considered (Figure 16).

Figure 16: 11 Sample Measured Rja Histogram (inset: 1000 Monte Carlo Simulated Rja Histogram)

From the larger sample population, statistical properties such as average and standard deviation can be extracted with more confidence. Such an approach is good for 'filling in the gaps' of under sampled measurements. In this case the average Rja value is 20.77 K/W with a normalized standard deviation of +/- 1.4%. At 3 sigma, this represents +/-4.8% which is therefore the maximum fluctuation in Rja this very level 2 package production guarantees (using these binned royal blue LEDs and same MCPCB). Implication of such a variation at thermal level will of course have implication on the electrical and optical fluctuation of these LEDs. This will be the topic of some future work within the scope of Delphi4LED.

7. Summary

A methodology to correlate observed variations in measured structure functions with the physical objects along the heat flow path in an LED assembly that are responsible for the variation has been presented. Two approaches are detailed; one taking an analytical approach considering the calculated cumulative thermal capacitance of the objects, the other using simulation by taking a TDIM perturbation type approach. It is concluded that the variation is due to changes in the thermal resistance of objects, as opposed to their thermal capacitance. Once the objects responsible for the variations are identified, more informed assumptions are made as to the root cause of the variations.

Accommodating the measured variation when performing thermal modeling is done by taking a Monte Carlo approach. A 12 stage RC ladder topology is chosen and 1000 models are constructed by determining normally distributed Rth values of each stage, taking into account the correlated interdependency of each stage. The 1000 models are solved and form that more dependable thermal metrics, such as mean and standard deviation of Rja, are calculated, more dependably than extracting the same values from the limited set of 11 measurements.

Acknowledgments

The contribution of the European Union for supporting this work in the context of the H2020 ECSEL Joint Under-taking programme (2016-2019) within the frames of the Delphi4LED project (grant agreement 692465) is acknowledged. Co-financing of the Delphi4LED project by the R&D funding organizations of the governments of the countries participating in this experiment through their national grant agreements is also acknowledged.

References

1. R. Bornoff, V. Hildenbrand, S. Lugten, G. Martin, C. Marty, A. Poppe, M. Rencz, W. Schilders, J. Yu, "Delphi4LED - From Measurements to Standardized Multi-Domain Compact Models of LEDs: a New European R&D Project for Predictive and Efficient Multi-domain Modeling and Simulation of LEDs at all Integration Levels Along the SSL Supply Chain", Therminic 2016, Budapest, Hungary
2. www.delphi4LED.org, retrieved January 2019
3. Poppe, J. Hegedűs, A. Szalai, R. Bornoff, J. Dyson, "Creating multi-port thermal network models of LED luminaires for application in system level multi-domain simulation using SPICE-like solvers", In: Proc. of the 32nd IEEE SEMI-THERM Symp., 14-17 March 2016, San Jose, USA, pp. 44-49, DOI: 10.1109/SEMI-THERM.2016.7458444
4. A. Poppe, "Simulation of LED Based Luminaires by Using Multi-Domain Compact Models of LEDs and Compact Thermal Models of their Thermal Environment", MICROELECTRONICS RELIABILITY 72:(5) pp. 65-74. (2017), http://dx.doi.org/10.1016/j.microrel.2017.03.039
5. www.mentor.com/products/mechanical/micred/t3ster/, retrieved January 2019
6. www.mentor.com/products/mechanical/micred/teraled/, retrieved January 2019
7. www.mentor.com/products/mechanical/micred/flotherm/, retrieved January 2019
8. JESD51-14, JEDEC. Standard "Transient Dual Interface Test Method for the Measurement of the Thermal Resistance Junction-To-Case of Semiconductor Devices with Heat Flow Through a Single Path". (2010)
9. Y. Zong, P-T. Chou, P. Dekker, R. Distl, K. Godo, P. Hanselaer, G. Heidel, J. Hulett, K. Oshima, A. Poppe, G. Sauter, M. Schneider, H. Shen, M.M. Sisto, A. Sperling, R. Young, W. Zhao, "Optical Measurement of High- Power LEDs", CIE 225:2017 (Technical Report) ISBN 978-3-902842-12-1, DOI: 10.25039/TR.225.2017
10. JEDEC JESD51-52 Standard "Guidelines for Combining CIE127:2007 Total Flux Measurements with Thermal Mea-surements of LEDs with Exposed Cooling Surface" (2012),
11. JEDEC JESD51-51 Standard "Implementation of the Electrical Test Method for the Measurement of Real Thermal Resistance and Impedance of Light-Emitting Diodes with Exposed Cooling" (2012), https://www.jedec.org/system/files/docs/JESD51-51.pdf https://www.jedec.org/system/files/docs/JESD51-52.pdf
12. D. Schweitzer, H. Pape, L. Chen, R. Kutscherauer and M. Walder, "Transient dual interface measurement — A new JEDEC standard for the measurement of the junction-to-case thermal resistance," 2011 27th Annual IEEE Semiconductor Thermal Measurement and Management Symposium, San Jose, CA, 2011, pp. 222-229.
13. Robin Bornoff, Thomas Mérelle, Josephine Sari, Alessandro Di Bucchianico, Gabor Farkas. Quantified Insights into LED Variability. 24rd International Workshop on Thermal Investigations of ICs and Systems (THERMINIC) 26-28 Sept. 2018
14. R. Bornoff, G. Farkas, L. Gaal, M. Rencz and A. Poppe, "LED 3D thermal model calibration against measurement," 2018 19th International Conference on Thermal, Mechanical and Multi-Physics Simulation and Experiments in Microelectronics and Microsystems (EuroSimE), Toulouse, 2018, pp. 1-7.
15. J. Sari, T. Mérelle, A. Di Bucchianico and D. Breton, "Delphi4LED: LED measurements and variability analysis," 2017 23rd International Workshop on Thermal Investigations of ICs and Systems (THERMINIC), Amsterdam, 2017, pp. 1-6.
16. T. Mérelle, R. Bornoff, G. Onushkin, L. Gaal, G. Farkas, A. Poppe, G. Hantos, B. Bataillou, J.K. Sari, "Modeling and Quantifying LED Variability," 8th International LED professional Symposium +Expo, 2018, Bregenz.

Accurate Thermal Transient Measurements Interpretation of Monochromatic LEDs

Alexeev Anton[1], Genevieve Martin[2], Grigory Onushkin[2], Jean-Paul Linnartz[2]
[1]Eindhoven University of Technology 5612AZ Eindhoven
a.alexeev@tue.nl
[2]Signify
High Tech Campus 48 5656AE Eindhoven

Abstract

In this work we determine the factors that influence the accuracy of LEDs thermal transient analysis, in particular, the leakage of heat into the dome material and the parasitic generation of heat on the reflector cup surface due to the light reflection losses are considered. Our analysis indicates a significant impact of these two factors on interpretation of the thermal transient measurements of mid-power LEDs. The paper demonstrates the significance of the physical phenomena behind these factors for creation of LEDs finite-element thermal models. We quantify the inaccuracies in the thermal structure functions associated with these effects. We determine possible inaccuracies of finite-element models parameters calibration by thermal structure function or transient thermal response alignment if these factors are not accounted for. We show a substantial impact of the parasitic heat losses on the LEDs' reflector cup surface on the results of thermal transient for LEDs with high internal quantum efficiency encapsulated in packages with low light extraction efficiency.

Keywords

LED, Secondary Heat Path, Dynamic Thermal Compact Model, Reliability, Thermal Transient Testing, Finite Element Analysis.

Nomenclature

FEA	Finite-Element Analysis
DAL	Die Attach Layer
DOE	Design of Experiments
DTCM	Dynamic Compact Thermal Model
EQE	External Quantum Efficiency
IQE	Internal Quantum Efficiency
IC	Integrated Circuit
LED	Light Emitting Diode
LEE	Light Extraction Efficiency
MP	Mid-Power
MOR	Model Order Reduction
PCB	Printed Circuit Board
SF	Structure Function
P_h	Dissipated Thermal Power
C_{th}	Thermal Capacitance
R_{th}	Thermal Resistance
T_j	Junction Temperature

1. Introduction

The design of modern lighting systems heavily relies on digital modeling and simulation of the system's components during the development phase. The designers execute a vast amount of numerical DOE scenarios prior to prototype creation to ensure high performance and reliability of the end product. This allows manufacturers to significantly decrease the time to market and reduce the associated development costs. However, this approach requires reliable multi-domain modeling of the components used. The key components of the modern lighting systems are LEDs. Our Delphi4LED European project aims at developing a standardized method to create multi-domain (thermal, electrical, and optical) compact models from the measurement data [1].

A major part of all LEDs failure modes are temperature-related [2]. Therefore, lighting systems design and reliability prediction critically relies on the ability to predict the thermal performance of the LEDs' packages inner structures such as DAL, cup reflector, solder, etc. In fact, accurate thermal modeling of LEDs' packages accelerates and advances design solutions and it enhances performance of the lighting systems [3].

DTCMs of LEDs provide high accuracy and require low computational resources which makes them a superior choice for performing computationally heavy numerical DOEs.

A common way of LEDs DTCM generation is thermal transient analysis proposed by V. Szekely [4]. The method enables the derivation of compact models for IC devices using the transient thermal response as input. Since this method yields a one-dimensional singular heat source DCTM, it has a number of drawbacks. The presence of multiple heat paths and heat sources in a real LED may affect the accuracy of the approach. Indeed, the secondary capacitive heat path associated with the heat propagation through the dome has been shown to affect the accuracy of this method [5]. In this paper, we deepen the understanding of the impact of the secondary heat path and we investigate the influence of parasitic heat generation on the LED's cup surface (See Figure 1) on the results produced by this method.

Another promising method of DTCM generation is MOR [6]. It enables the derivation of a DTCM from a detailed thermal finite-element model. To ensure the accuracy of the derived DTCM, the detailed finite-element thermal model has to represent the original LED steady-state and transient thermal behavior precisely. Thus, both the geometry and the material properties used in the detailed finite-element thermal model have to be calibrated prior to applying MOR.

The geometrical features of the LED's package can be obtained directly by performing either a cross-section or x-ray scans. The thermal properties of the material in the LEDs' packages are significantly harder to measure explicitly due to the small size of these packages. Therefore, firstly a rough initial estimation of the thermal properties is done based on approximate knowledge of the materials in the LEDs' packages structures. Next, these thermal properties are calibrated to enhance the accuracy of thermal model. Measurement data of the T_j transient response is often used for the calibration [7].

Nevertheless, calibration of finite-element models by T_j as a single-valued parameter has significant limitations. First of all, a three-dimensional LED package dynamic thermal response collapses to a singular time-dependent (yet, effectively only a one-dimensional) T_j response. Thus, interactions among multiple heat paths can introduce calibration errors. Next, the presence of multiple heat sources inside of an LED package can significantly affect the accuracy of the calibration since often only a singular pn junction heat source is considered to simulate the heat generation [8]. Yet the pn junction may not be the only significant heat source. The reflector cup surface and other LEDs package structures, which are exposed to the high intensity light flux, may cause substantial additional parasitic heat losses. Although these are often neglected, this paper demonstrates the importance of accounting for these losses for modern LEDs thermal models creation and calibration.

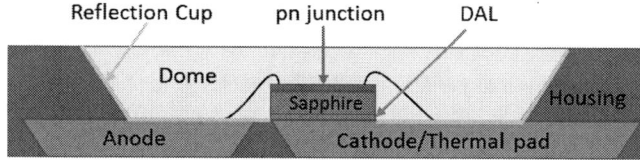

Figure 1. LEDs schematic cross-section. Major considered structures are designated.

In this work, we use a mockup of a MP LED package and we consider a range of possible LEE coefficients without an explicit ray tracing analysis, to demonstrate generically the impact of the considered phenomena.

2. Methods

2.1. Thermal Transient Analysis

Thermal transient testing is a well-known method to characterize IC packages. The method derives the characteristic time constant spectrum of devices' transient T_j response to a power step and maps it to exponential functions. T_j is typically obtained by sampling the forward voltage of the device under test. The dependency of the forward voltage on T_j can be found by device calibration with a temperature controlled heat sink. The method requires additional procedures to characterize the LED. LEDs emit a significant fraction of the applied electrical power as light, thus the total power dissipated by the LED's pn junction is considerably smaller than the applied electrical power. The optical power have to be determined with a radiant power measurement, usually performed by an optical integration sphere. The detailed procedure of LEDs thermal transient testing is described in details in the JESD51-5x series of standard [9]. The thermal transient analysis can also be used for analysis of the transient T_j response obtained with finite-element transient simulations.

An SF is commonly used to represent thermal transient measurements. A SF is a one-dimensional heat-flow map of the device under test. An SF can be graphically plotted as the

devices cumulative C_{th} versus its cumulative R_{th} along the heat path starting from the pn junction. Examples of SFs are presented in Figure 3 and the following figures. Elements with high partial R_{th} and low C_{th} (e.g. DAL) are reflected on the SF plots as shelfs, elements with low partial R_{th} and high C_{th} (e.g. thermal pad) as steps. This simplifies the interpretation of an SF and it maps the characteristic features of a SF to certain elements of the heat path of the device under test. In this work, the impact of the multiple heat path and multiple heat source presence is determined by SFs analysis and comparison.

The finite-element model calibration can conveniently be done by aligning the transient analysis derived from SFs. This approach enables an advanced calibration method since any misalignment of the SFs can be mapped to a particular part of the thermal path. This emphasizes the importance of the objective of this paper, namely to investigate the secondary heat path and the secondary heat source impact on the transient response analysis results and their interpretation. The paper determines and characterizes the SF inaccuracies caused by these factors.

2.2. Finite-Element Thermal Modeling

We use a finite-element model of a typical mid-power LED package mockup mounted on a metal core printed circuit board and we analyze only general trends of the SF distortion caused by the considered sources of the inaccuracy. Therefore, the precise geometrical details of the package are omitted since their effect on the SFs is minor and unique for each LED package.

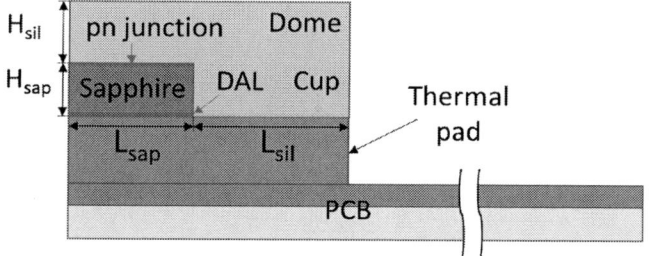

Figure 2. Axisymmetric thermal model of the LED mounted on a PCB used for the finite-element simulations.

We use an axisymmetric simplified LED package similar to a typical MP 3030 LED. Thus, we employ the numerical two-dimensional finite-element model presented in Figure 2. We simulate the transient T_j response to a power step. A constant temperature boundary condition is set to the bottom of the PCB, which imitates a controlled heatsink isothermal surface. Adiabatic thermal boundary conditions are set to all the other outer model surfaces. Convection and radiation heat transfers are omitted as the fraction of the heat leaving the LED via these mechanisms is negligible compared to the main heat path heat flow [5]. Two heat sources, both with a uniform heat generation profile, are considered: the pn junction and the thermal pad surface exposed to the dome. We are not modeling the LEDs plastic housing since it is located far from the pn junction and has low thermal conductivity, thus negligible effect on the T_j response. Nevertheless, we account for the optical losses on the side wall of the LED as if the plastic housing was there. These losses are added to the thermal pad surface heat losses and evenly distributed across it. We assume that material properties do not depend on the temperature. We do not simulate precise geometry of the silicone dome but model it as a bulk silicone cylinder of various heights. The thermal properties of sapphire crystal, silicone dome, copper thermal pad and FR4 materials that can be found in open literature were used. L_{sil} is set to 1.2mm, L_{sap} to 0.4mm, H_{sap} to 0.19mm. We sweep H_{sil} between 0 and 0.3mm.

2.3. Optical Power Distribution Modeling

The fraction of the light initially emitted by the crystal that escapes the LED is described by the LEE_{dome} coefficient. The efficiency of light extraction from the crystal to the silicone dome without considering light re-entries is described by the LEE_{chip} coefficient. The total thermal power P_h dissipated inside of an LED consists of the power P_{h_cup} dissipated on the cup surface and the power P_{h_pn} dissipated inside of the pn junction. The thermal power dissipated on the walls is dependent on the optical power emitted by the pn junction, the cup geometry and the materials used.

Nevertheless, in a real LED a part of the light that initially leaves the sapphire crystal can be reflected back and can be absorbed by the crystal or the active area. The amount of heat dissipated on the cup surface (inside of the sapphire crystal) due to the optical losses of the light that initially left the crystal is proportional to the number of reflections from the cup surface N_{cup} (crystal transitions N_{cry}) and the corresponding absorption coefficients A_{cup} and A_{cry}. We assume that N_{cup} and N_{cry} are proportional to the surface areas of the cup S_{cup} and the sapphire crystal S_{cry} that are in contact with the dome. Thus, the amount of the energy absorbed is proportional to the product of corresponding surfaces and the absorption coefficients. The absorption coefficients are directly related to the reflection coefficient of the walls R_{cup} and the transition coefficient of the crystal T_{cry} as $A_{cup} = 1 - R_{cup}$ and $A_{cry} = 1 - T_{cry}$. We assume T_{cry} to be equal to LEE_{cry}. The R_{cup} is 0.93 for a silver surface and monochromatic blue light. We use characteristic surface area of the cup that is correspondent to a typical 3030 MP single die LED. This yields Formula 1 defining the ratio between the thermal powers.

$$\frac{P_{h\_cup}}{P_h} \cong \frac{IQE \times LEE_{chip} \times [1-LEE_{dome}]}{1-IQE \times LEE_{chip} \times LEE_{dome}} \times \frac{S_{cup} \times (1-R_{cup})}{S_{cry} \times (1-LEE_{chip}) + S_{cup} \times (1-R_{cup})} \tag{1}$$

We implicitly consider various cup and silicone dome optical configurations (e.g. dome curvature, cup walls height and angle etc.) by accounting for an expected range of values correspondent to the LEE_{dome} coefficient normalized with respect to the sapphire-silicone light extraction (light emitted into the encapsulant for the first time). We use results by N. T. Tran [10] to identify this range. The results show evidence that the LEE_{dome} coefficient lies in the range from 65% to 97% for conventional designs of MP LEDs packages. The light extraction efficiency from the sapphire crystal to the dome LEE_{chip} can reach 85% for double-side textured-LED designs [11]. The characteristic IQE can be as high as 86% for modern blue LEDs at room temperatures [12] and continue to increase.

According to Formula 1 and the determined range of the possible LEE parameters P_{h_cup} can rich up to 35% of P_h. Such high ratio is characteristic for LEDs packages with low LEE_{dome} designs (e.g. flat domes) and chips with high IQE (e.g. modern blue LEDs) Yet, the P_{h_cup} to P_h ratio can also be close to zero for LEDs with high LEE_{dome} and low IQE. The power ratio almost linearly scales down to zero with decrease of the IQE. In order to determine the exact power ratio ray tracing analysis, total light flux and IQE measurements have to be done for a specific LED of interest.

3. Results and Discussion

3.1. Silicone Dome Presence Influence

Firstly, we perform a transient FEA simulation activating a singular heat source in the finite-element model. We set all the thermal power to be generated exclusively inside of the pn junction of the LED chip. No parasitic heat generation on the cup surface is simulated. We investigate the influence of

silicone dome height H_{sil} on the SF distortion separate from the other factors.

SFs correspondent to the MP LEDs FEA simulation results are shown in Figure 3 and 4. Number of silicone dome heights in range from 0 to 3e-4 m are considered all the other parameters of the model are kept constant.

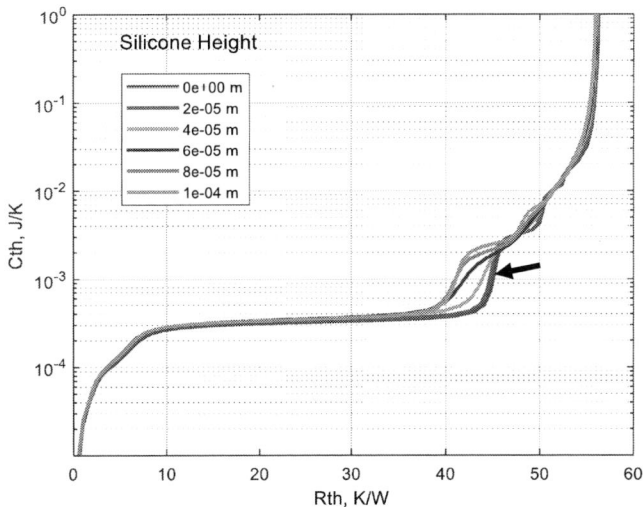

Figure 3. Effect of the silicone dome presence on a characteristic MP LEDs SF. For 0 m<H_{sil}<1e-4 m. Thermal pad step of the H_{sil}=0 configuration (marked with an arrow) determines the actual junction-to-thermal pad thermal resistance of 45 K/W.

Figure 3 shows a substantial influence of the silicone dome height on the SFs. The SF distortion significantly increases until the silicone height on top of the sapphire chip not reaches 1e-4m. Further increase of the silicone height does

9

not considerably contribute to the SF distortion. The DAL-thermal pad step is the region of the SF the most affected by the silicone height increase: the increase of the silicone height leads to a decrease of the pn junction to the thermal pad R_{th} value derived with the transient thermal analysis. This effect was already described in [5]. The distortion of the R_{th} value is associated with heat storage inside of the silicone dome. The thicker the dome, the more heat can be stored inside of it. Nevertheless, thermal conductivity of silicone is as low as 0.2 W/m·K. Thus, the distortion saturates when the dome thickness exceeds certain value since the upper layer of the silicone dome cannot effectively influence the T_j response due to the significant thermal resistance of the silicone located below them.

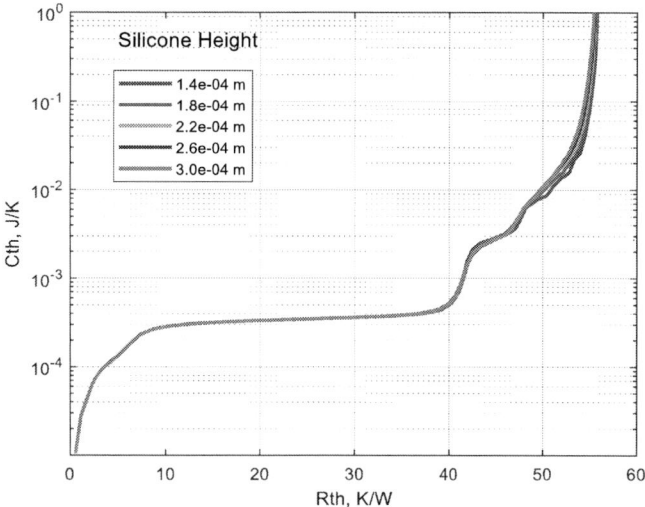

Figure 4. Effect of the silicone dome presence on a characteristic MP LEDs SF. For 1.4e-4m<H_{sil}<3e- m. No significant variations of the extracted SFs are noticeable. The SF distortion is saturated.

3.2. Optical Heat losses Influence.

To explore the impact of parasitic optical heat losses on the transient analysis results, we employ our finite-element model with two active heat sources. We consider heat generation inside the pn junction and on the thermal pad surface. We fix the H_{sil} at 2e-4m and sweep P_{h_cup} to P_h ratio in the range determined in the previous section (0 to 0.35). The results are illustrated in Figure 5.

We observe a strong dependence of the SFs' on the power split ratio. Junction to ambient R_{th} value significantly decreases with the increase of P_{h_cup}. Moreover, both the C_{th} and R_{th} values of the SFs along all the heat path scale with the increase of the P_{h_cup} fraction. This inspires us to assume that the parasitic heat generation on the cup surface may significantly affect the results of the transient analysis and lead to an error of LEDs thermal parameters extraction by the thermal transient analysis if the parasitic heat generation is not accounted for.

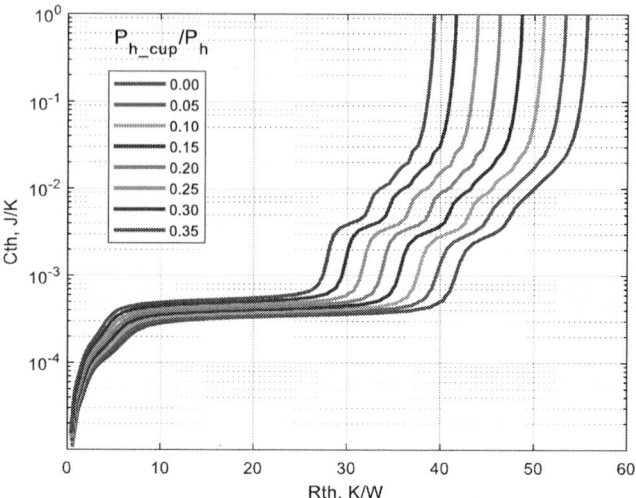

Figure 5. Influence of the parasitic heat dissipation on the cup surface on the SF. Total thermal power P_h is used for the transient analysis.

The scaling of the SFs occurring with the change of the P_{h_cup} to P_h ratio inspires us to use P_{h_pn} value instead of P_h for the transient analysis. Normalization of the thermal power with this value leads to a better alignment of the SFs plots up to the characteristic cathode SF step and significantly decreases the spread of the derived junction to ambient R_{th} values. Indeed, a major part of a typical MP lateral LED R_{th} is related to the DAL and sapphire crystal partial thermal resistances. The heat flux through them is determined by the heat generation in the active area. Thus, an error of the determination of the power split between the P_{h_cup} and P_{h_pn} may lead to a wrong determination of the sapphire crystal and DAL R_{th} with the transient analysis method. Nevertheless, after the power adjustment the SFs' plots are still not aligned after the thermal pad step. This is caused by the fact that the parasitic thermal power is being dissipated on the thermal pad surface. This additional power is not accounted for in the considered case.

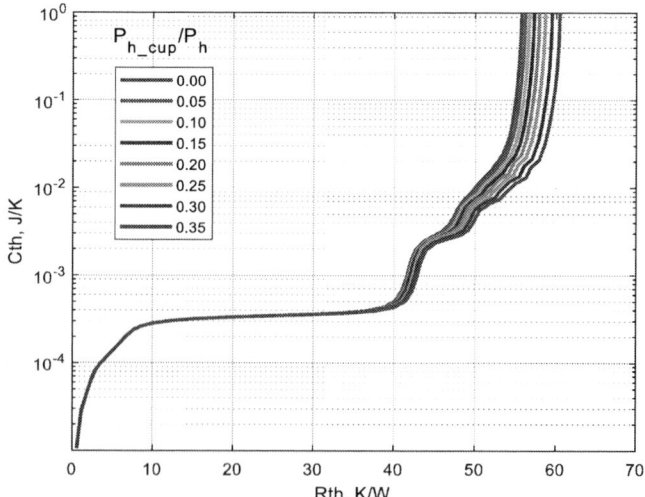

Figure 6. Influence of the parasitic heat dissipation on the cup surface on the SF. Thermal power dissipated at the junction P_h-P_{h_cup} is used for the transient analysis.

Now, knowing how the characteristic SF distortion depends on the power split ratio, we can relate it back to particular LEDs packages with certain LEE_{dome} and IQE values. E.g. modern blue MP LEDs have an IQE of approximately 70% under the standard bias current. The P_{h_cup} to P_h ratio in this case lies in the range of 0.05 to 0.2 depending on the EQE_{dome} [10]. As can be seen in Figure 5, these numbers lead to the junction-to-thermal pad R_{th} error of 6% to 25%. Under standard operating conditions IQE of monochromatic green LEDs chips is approximately 40%. This yields a P_{h_cup}-to-P_h ratio range of 0.02 to 0.1, which translates to 1% to 12% of junction-to-thermal pad R_{th} error. The SFs of the less efficient LEDs experience significantly smaller distortion due to the considerably lower IQE which decreases the P_{h_cup} to P_h ratio and reduces the parasitic effects.

4. Conclusions

The characteristic relative error of the SF derived junction-to-thermal pad R_{th} values associated with the silicone dome thermal capacitive effect is shown to be as large as approximately 10% for mid-power LEDs. This number is in a good agreement with the previously obtained experimental measurements and the numerical simulations results. We found a saturation of the SF error, caused by the dome thermal capacitance for the silicone dome thickness H_{sil} higher than 0.1 mm.

The SF distortion caused by the optical heat losses on the reflector cup surface is demonstrated for the first time in literature. It impacts the sapphire and the DAL SF regions. The error of the SF derived junction-to-thermal pad R_{th} value associated with the optical heat losses can reach up to 25% for highly efficient LEDs encapsulated in packages with low LEE_{dome} under normal bias current. The error depends on the IQE, LEE_{chip}, LEE_{dome} coefficients and on the package geometry. The estimations presented in the paper are done for a range of possible IQE and LEE parameters characteristic for MP LEDs. Our results presented here emphasize the importance of performing the ray tracing simulations for proper calibration of the thermal finite-element models of LEDs with high IQE.

Neglecting parasitic heat losses on the reflector cup surface during finite-element thermal modelling can lead to a significant estimation inaccuracy of the main heat path thermal properties during calibration with the measured transient response data.

The distortion of the SFs associated with the heat propagation to the dome is automatically compensated if the dome is physically modeled. Nevertheless, even if a DTCM is to be derived directly from a measurement results with help of thermal transient analysis, this error has to be also accounted for.

5. Acknowledgment

The contribution of European Union is acknowledged for supporting the study in the context of the ECSEL Joint Undertaking program #692465 (2016–2019). Additional inforation is available on: www.DELPHI4LED.eu.

6. References

[1] R. Bornoff *et al.*, "Delphi4LED - From measurements to standardized multi-domain compact models of LED: A new European R&D project for predictive and efficient multi-domain modeling and simulation of LEDs at all integration levels along the SSL supply chain," *THERMINIC 2016 - 22nd Int. Work. Therm. Investig. ICs Syst.*, pp. 174–180, 2016.

[2] W. D. van Driel, F. Xuejun, and Z. Guoqi, *Solid State Lighting Reliability*. New York, NY: Springer New York, 2013.

[3] A. Alexeev *et al.*, "Requirements specification for multi-domain LED compact model development in Delphi4LED," in *2017 18th International Conference on Thermal, Mechanical and Multi-Physics Simulation and Experiments in Microelectronics and Microsystems (EuroSimE)*, 2017, pp. 1–8.

[4] T. Van Bien and V. Szekely, "Fine structure of heat flow path in semiconductor devices: A measurement and identification method," *Solid. State. Electron.*, vol. 31, no. 9, pp. 1363–1368, 1988.

[5] A. Alexeev, G. Martin, and G. Onushkin, "Multiple heat path dynamic thermal compact modeling for silicone encapsulated LEDs," *Microelectron. Reliab.*, vol. 87, no. May, pp. 89–96, Aug. 2018.

[6] S. Lungten, R. Bornoff, J. Dyson, J. M. L. Maubach, W. H. A. Schilders, and M. Warner, "Dynamic Compact Thermal Model Extraction for LED Packages Using Model Order Reduction Techniques," vol. m, no. September, pp. 1–6, 2017.

[7] R. Bornoff, G. Farkas, L. Gaal, M. Rencz, and A. Poppe, "LED 3D Thermal Model Calibration against Measurement," pp. 1–7, 2018.

[8] A. Poppe, "Multi-domain compact modeling of LEDs: An overview of models and experimental data," *Microelectronics J.*, vol. 46, no. 12, pp. 1138–1151, 2015.

[9] A. Poppe, "Testing of Power LEDs : The Latest Thermal Testing Standards from JEDEC," *Electron. Cool. Mag.*, no. September 2013, 2013.

[10] N. T. Tran and F. G. Shi, "LED package design for high optical efficiency and low viewing angle," *2007 Int. Microsystems, Packag. Assem. Circuits Technol.*, pp. 10–13, 2007.

[11] T. Chen *et al.*, "Improvement in Light Extraction Efficiency of High Brightness InGaN-Based Light Emitting Diodes," vol. 7216, pp. 1–10.

[12] I. E. Titkov *et al.*, "Temperature-Dependent Internal Quantum Efficiency of Blue High-Brightness Light-Emitting Diodes," *IEEE J. Quantum Electron.*, vol. 50, no. 11, pp. 911–920, Nov. 2014.

[13] A. Alexeev, G. Martin, and V. Hildenbrand, "Structure function analysis and thermal compact model development of a mid-power LED," in *2017 33rd Thermal Measurement, Modeling & Management Symposium (SEMI-THERM)*, 2017, pp. 283–289.

Implementation of a Multi-domain LED Model and its Application for Optimized LED Luminaire Design

János Hegedüs[1], Gusztáv Hantos[1], Robin Bornoff[2], Márta Rencz[1,3], András Poppe[1,3] *

[1] Budapest University of Technology and Economics, Budapest, Hungary
[2] Mentor, a Siemens Business, Hampton Court, UK
[3] Mentor, a Siemens Business, Budapest, Hungary
*E-mail: poppe@eet.bme.hu, andras_poppe@mentor.com

Abstract

In the Industry 4.0 era digitalization of design and manufacturing processes is taking place. The aim of the Delphi4LED H2020 ECSEL R&D project of the EU is to trigger such a transition in the solid-state lighting industry by developing testing and modelling methodologies aimed at multi-domain characterization of LED based products at different levels of integration along the SSL supply chain. Key to the support of the fully digitalized LED luminaire design process is the appropriate implementation of a chip level LED multi-domain model that can be used in design tools which are widely available both for SMEs and big international companies. In this paper, we present the latest implementation of BME's chip level multi-domain LED model and its application in an spreadsheet based simulation tool. The model has been successfully tested through the first demonstrator design of the Delphi4LED project.

Keywords

LED mult-domain modelling, compact modelling, SPICE models, spreadsheet application

1. Introduction

In terms of modelling and simulation, until recently, there was a gap between LED chip design/manufacturing and lighting design. The link between the LED light source and the lighting task is the *luminaire*. Luminaire design for traditional light sources required mainly optics design only. With LEDs, as semiconductor devices sensitive to heat, thermal aspects also have been considered in LED luminaire design. Tools widespread in design of classical electronics cooling such as CFD simulators have been natural choices to support LED luminaire design as well. Still, due to the lack of multi-domain compact models of LED packages, optimized LED luminaire design considering all the thermal, electrical and optical domains of LED operation simultaneously was not yet possible.

The Delphi4LED European R&D project [1] aimed to bridge this gap through the development of testing, modelling and simulation methodologies and tools aimed at LED package, module and luminaire level. In terms of modelling, a hierarchical, modular approach is used in order to create a system level compact model of a complete LED luminaire that can be used for the luminaire's multi-domain optimization.

Inputs to such an optimization are the overall lighting requirements such as target minimal hot lumens, maximal

input electrical power, minimal driver and optics efficiency. As a result of the optimization process, luminaire designers can have the best selection of LED packages and their physical arrangement, choice of thermal management solutions, etc. that assure achieving the lighting design specifications without unnecessary safety margins in the design. In the bottom-up modular approach of modelling the basis is the multi-domain model of LED chips [2] suitable for Spice-like simulations, followed by compact thermal models of the LED packages, modules and finally, luminaires [3], [4]. In this paper we report about the latest achievements in terms of chip level multi-domain modelling of LEDs and the implementations of such LED models in software tools widely available to designers, such as generic SPICE circuit simulators of Excel spreadsheet applications.

2. Methodology

The recent implementation is based on an existing and well detailed Spice-compatible multi domain LED model [2]. The baseline has already been utilized in system level luminaire simulations [3] validated by both field and laboratory measurements [5], [6], [7]. The baseline model [2] overestimated the LEDs' efficiency at lower currents and temperatures. This has been mitigated in two ways [8], [9]: on the one hand, as seen in Figure 1, an extra resistor has been added to the optical branch of the original model to account for optical losses (resolving the efficiency overestimation issue), on the other hand, a modified, improved parameter extraction approach further contributed to the accuracy of the multi-domain LED chip model, resulting in an excellent fit over two decades (from ~10 mA up to ~1 A) of the forward current at all relevant junction temperatures (between 30 $^\circ$C and 110 $^\circ$C).

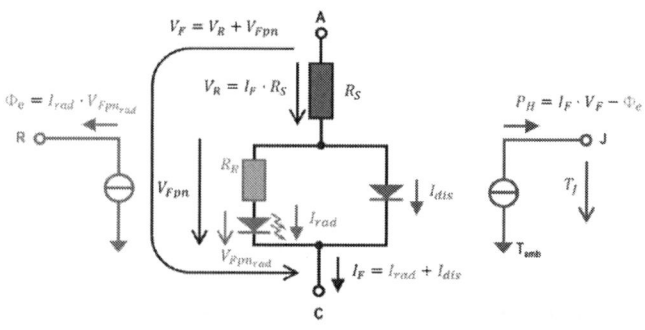

Figure 1: Topology of the improved LED model [8], [9].

2.1. Voltage controlled vs. current controlled model

Our baseline model [1] was originally aimed at implementation as built-in or macro-models in SPICE-like circuit simulators with full electro-thermal capabilities such as Mentor's ELDO [10]. In this original approach, the multi-domain LED model was a so called *voltage driven model*, in compliance with the so called *nodal voltage approach* used in Spice-like circuit simulators to solve Kirchhoff's equations. In our present approach, this has been "reversed", i.e. the forward voltage is calculated from the applied steady forward current, reflecting how LEDs are used and tested in reality, i.e. our present version of the multi-domain LED model is a so called *current controlled* model. In this approach the model's nonlinearity is much weaker (logarithmic dependence on the forward current) than in the traditional, voltage controlled approach (exponential dependence on the applied forward voltage). The input of such a chip level multi-domain LED model are the I_F forward current and the T_J junction temperature (as indicated in which in daily engineering practice are also among the major specs of an LED based luminaire. As output, the model provides the LEDs' V_F forward voltage, Φ_e radiant flux (aka emitted optical power, also denoted by P_{opt}), Φ_V luminous flux and P_H power dissipation. This approach to chip level multi-domain modelling allows embedding our new model into any thermal simulator for a *relaxation type* coupled electro-thermal-optical simulation of complex LED applications. From the P_H dissipation of the LED chips the 3D thermal solver provides the updated values of the T_J LED junction temperatures, as illustrated in Figure 2.

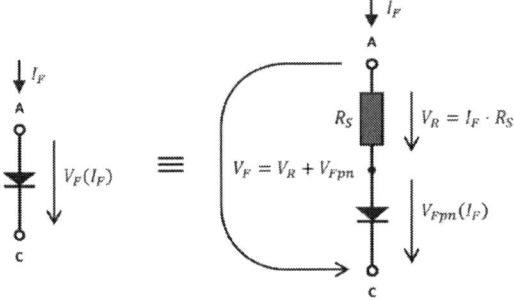

Figure 2: Application of the reformulated chip level multi-domain LED model in a relaxation type implementation of an electro-thermal-optical solver called *LED Luminaire Design Calculator* [11].

Figure 3: An LED as a single semiconductor diode, with its electrical properties I_F forward current and V_F forward voltage between its anode and cathode, the R_S electrical series resistance and the V_{Fpn} voltage drop on the "internal junction".

2.2. Modelling for generic Spice solvers

Besides reformulating the previous voltage driven LED model to a current driven one, a second change in our modelling philosophy was also applied. In order to allow implementation of the model for generic SPICE solvers without electro-thermal simulation capabilities, first we model the LED as a generic semiconductor diode, relying solely on the classical Shockley model of pn-junctions, completed with an electrical series resistance, see Figure 3. We also call this model as the "main diode".

Figure 4: Temperature dependence of the forward voltage of a constant DC forward current driven diode.

Since the applied forward current of the LED is a constant input value of the model, the temperature dependent shift of the forward voltage (see Figure 4) should be modelled to represent the electro-thermal operation of the LED. This voltage shift is known as TSP (temperature sensitive parameter) in JEDEC's semiconductor device thermal testing standard JESD51-1 and in its extension to LEDs, the JESD51-51 standard. In thermal testing based on these standards this temperature voltage shift is modelled by the concept of the K-factor. For a generic SPICE circuit simulator (with no electro-thermal simulation capabilities) this voltage shift can be represented by temperature dependent voltage generator as shown in Figure 5.

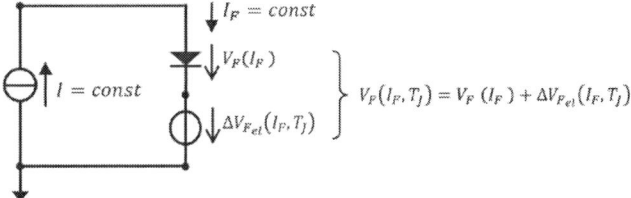

Figure 5: Modelling the effect of self-heating of a constant DC forward current driven diode in an electrical-only SPICE simulator.

As also known from JEDEC's LED thermal testing standard JESD51-51 [12], the K-factor depends on the applied forward current. Our recent measurement [13] revealed that in the relevant forward current and junction temperature ranges relevant for today's high power LEDs, the K-factor no longer can be considered as a temperature independent constant value, therefore in the new implementation of our multi-domain LED model the $\Delta V_{Fel}(I_F, T_J)$ generator is represented by the following equation:

$$\Delta V_{F_{el}} = (a_{el} \cdot I_F^2 + b_{el} \cdot I_F + c_{el}) \cdot (T_J^2 - T_{ref}^2) + \\ (d_{el} \cdot I_F^2 + e_{el} \cdot I_F + f_{el}) \cdot (T_J - T_{ref}). \quad (1)$$

where T_{ref} is an arbitrarily chosen reference temperature for which the parameters of the generic SPICE electrical-only diode model are valid. the necessary coefficients can be identified from advanced LED K-factor calibration results. Note, that the above relationship is valid for the entire light emitting diode as a whole, as it is also physically tested. Thus, as a major change in our modelling philosophy is the total I_F forward current is no longer calculated as the sum of the separately modelled I_{rad} and I_{dis} components (as shown in Figure 1). What we kept from the previous modelling concept is that the radiative part of the LED is also described by the Shockley-model.

Figure 6 illustrates our present approach: the entire LED is modelled as described above (Figure 6a) and with the radiative part described in a similar manner (Figure 6b). This re-formulation of the multi-domain LED model allows its implementation both as a set of model equations coded in Visual Basic macros, allowing to use the model in a spreadsheet application (see Figure 2) or as subcircuit macros in standard SPICE solvers without electro-thermal simulation capabilities (see e.g. Figure 7). Key to this is using a temperature independent diode model completed with the temperature dependent ΔV_{Fel} and ΔV_{Frad} generators indicated in Figure 6. Note in Figure 7 that in case of the voltage offset generators Eq. (1) is directly implemented in SPICE.

2.3. Light emission and heat dissipation

The pn-junction in the modified LED model (see the electrical schematics in Figure 6) is still split up to a virtual dissipative and a radiative branch (Figure 6b) but in this case the forward current is not divided. Instead, the radiative branch inherits the voltage of the electrical pn-junction.

The connection between the two parts of the model is realized through the $I_{rad} = P_{opt}/V_{Fpn}$ relationship where P_{opt} denotes the emitted optical power of the LED (not indicated in Figure 6), which is also denoted by Φ_e.

The P_H heating power of the LED (as input for thermal modelling) is calculated as defined in JEDEC's LED thermal testing standard JESD51-51: $P_H = I_F \cdot V_F - P_{opt}$.

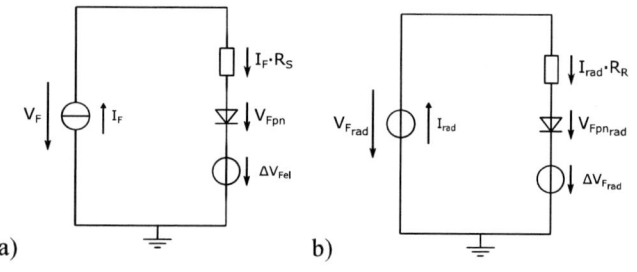

a)   b)

Figure 6: Schematics of a) the overall electrical model of an LED chip, and b) the model of the optical part of the multi-domain LED model.

The luminous flux (aka "hot lumens") is calculated directly from the P_{opt} emitted optical power (not shown in Figure 6), using the following empirical model for the *luminous efficacy of source of radiation*:

$$K(I_F, T_J) = (a_{Kap} \cdot T_J^2 + b_{Kap} \cdot T_J + c_{Kap}) \cdot I_F^2$$
$$+ (d_{Kap} \cdot T_J^2 + e_{Kap} \cdot T_J + f_{Kap}) \cdot I_F \quad (2)$$
$$+ (g_{Kap} \cdot T_J^2 + h_{Kap} \cdot T_J + i_{Kap}).$$

With this, the luminous flux is calculated as
$$\Phi_V = P_{opt} \cdot K(I_F, T_J). \quad (3)$$
All together the re-formulated chip level multi-domain LED model has 20 parameters and the additional 8 coefficients of Eq. (2) are used for the hot-lumen calculations.

Symbol	Meaning
T_{ref}	arbitrary chosen reference value of the junction temperature (in Kelvins); also known as T_{nom}
U_T	the thermal voltage calculated for T_{ref}
I_0	Standard Spice electrical diode model parameter IS known as *saturation current*, also known as the current coefficient of the Shockley-equation of diodes (for the T_{ref} reference temperature, for the main diode)
m	Standard Spice electrical diode model parameter N known as *ideality factor* or *emission coefficient* (for the main diode)
R_S	Standard Spice electrical diode model parameter RS known as the *electrical series resistance* (for the T_{ref} reference temperature, for the main diode)
$I_{0_{rad}}$	Standard Spice electrical diode model parameter IS (for the T_{ref} reference temperature, for the radiative branch of the LED model)
m_{rad}	Standard Spice electrical diode model parameter N (for the radiative branch of the LED model)
R_R	Standard Spice electrical diode model parameter RS (for the T_{ref} reference temperature, for the radiative branch of the LED model)
a_{el}	Coefficients of the model of the ΔV_{Fel} generator model for the main diode. These parameters are related to the K-factor of the LED; they can be identified from K-factor calibration curves obtained for different forward current values
b_{el}	
c_{el}	
d_{el}	
e_{el}	
f_{el}	
a_{rad}	Coefficients of the model of the ΔV_{Frad} generator model for the radiative branch
b_{rad}	
c_{rad}	
d_{rad}	
e_{rad}	
f_{rad}	
a_{Kap}	Coefficients of the model of the efficacy of source of radiation. Purely empirical values. (The efficacy of source of radiation is a property of the spectral power distribution of the emitted light of a light source.)
b_{Kap}	
c_{Kap}	
d_{Kap}	
e_{Kap}	
f_{Kap}	
g_{Kap}	
h_{Kap}	
i_{Kap}	

Table 1: Summary of the model parameters.

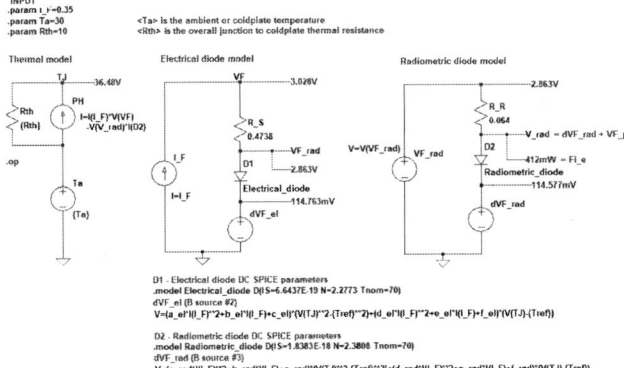

Figure 7: LT-Spice implementation of the re-formulated multi-domain LED chip model, completed with a single-R_{th} thermal model for testing purposes.

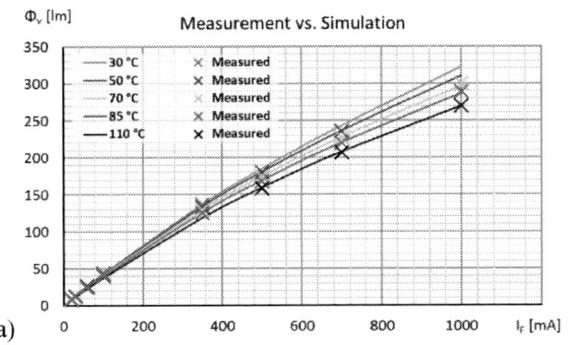

a)

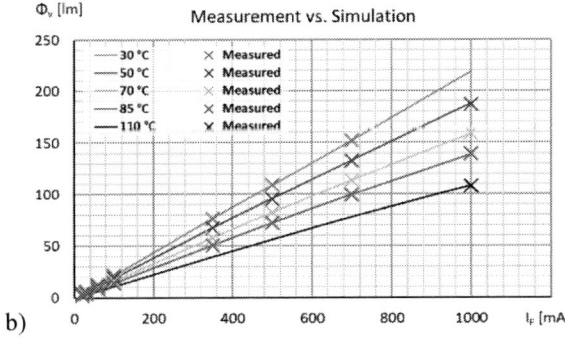

b)

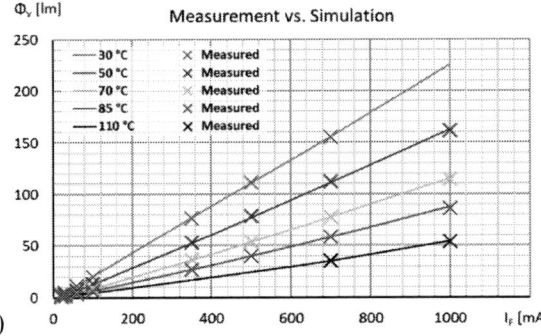

c)

Figure 8: Luminous flux as the function of forward current and junction temperature of a) phosphor converted white LED b) red LED c) amber LED.

The complete set of model parameters is summarized in Table 1. Figure 7 presents the LT-Spice implementation of the model, having the same set of model parameters as presented in Table 1.

3. The model applied to real LEDs

Within the Delphi4LED project a total number of 75 LED samples of 6 different types were characterized in 35 operating points. A global parameter fitting procedure was performed on all the individual measurement sets. Then two average models per each LED type were set up with the intent to model the LED type instead of the individual samples: one by averaging the resulted model parameters of the same LED type and another one by averaging the corresponding measurement results first and performing the fitting procedure afterwards.

Figure 8 compares the measured and modelled luminous flux values of a phosphor converted white, a red and an amber high power LED, as the function of forward current and junction temperature. The measurement results of the red and the amber LEDs also show significant temperature dependence, indicating the importance of the hot-lumen simulations. The diagrams are screen captures of an Excel spreadsheet application used for model parameter extraction. This application contains the Visual Basic macro version of the new LED model. The same set of Visual Basic macros are built into a spreadsheet application based *LED Luminaire Design Calculator* [11].

Based on our current experiences so far we can say that high dispersion of the ideality factor could spoil component-type modelling achieved by the parameter averaging method. Additional models were generated with the same LEDs but with reduced ideality factor deviation of the gained model parameters. This way more consistent model parameter sets could be achieved. Altogether, over 900 different models of the sample LEDs were set up in order to assist the further investigations on variability analysis that have also been started within the Delphi4LED project [14], [15], [16].

4. Demonstration of the use of the new LED model

In the Delphi4LED project, two types of "Industry 4.0"-like workflows have been proposed for implementation aimed at luminaire design [11]. One of the workflows, aimed at larger companies having sufficient resources to maintain specialized design teams and specialized, high-end commercial simulation software tools (Spice solvers, CFD tools) uses the Spice circuit macro implementation of our LED multi-domain model (Figure 7). In this workflow the chip level multi-domain LED model is used in automated optimization of the thermal design of the luminaires.

Another workflow is primarily aimed at designers working at SMEs with limited budget for advanced simulation tools and maintaining specialized design teams within their organization. This so-called SME workflow is based on the *Luminaire Design Calculator*, an Excel spreadsheet based application developed specifically for the Delphi4LED project, see Figure 9. The Visual Basic macro implementation of our multi-domain LED model with extracted set of model parameters (see Table 1) is part of this application along with a library of compact thermal models of different LED module substrate versions (with different substrate materials and LED layout arrangements) and simple compact thermal models of luminaires acting as heatsinks.

The *Luminaire Design Calculator* is equipped with a simple user interface, see Figure 9. The luminaire designer has to provide the major lighting **design goal** in terms of the *total emitted luminous flux of the luminaire* and has to provide the **design constraints** such as maximum allowed temperatures and maximum total electrical input power. The third group of input parameters describes the major properties of a foreseen luminaire **design** variant such as the total number of LEDs (in series electrical configuration) to be used in the luminaire, foreseen input electric current (the common I_F forward current of the LED string), driver and optics efficiencies and the choice of LED module substrate and luminaire heatsink. Once all these inputs are provided, the spreadsheet application creates the compact thermal model of the LEDs' 3D environment and calculates the LEDs' operating points corresponding to the specified ambient temperature and forward current through a relaxation type iterative solution process as illustrated in Figure 2.

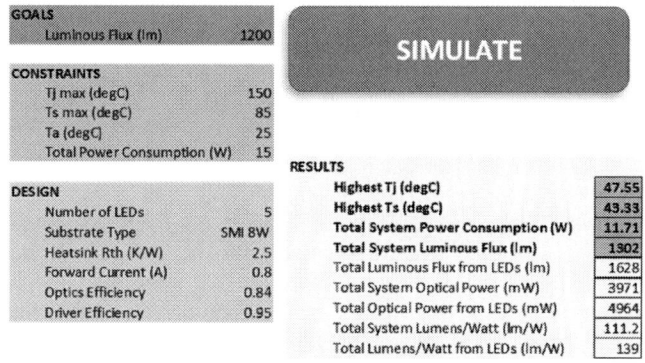

Figure 9: The user interface of the Luminaire Design Calculator Excel spreadsheet application with the design input settings and calculated results of the final, optimized version of the first Delphi4LED demonstrator design.

The LED *Luminaire Design Calculator* and a SPICE solver, both using the corresponding implementation variant of the presented LED model (with the extracted parameter set of Cree's XPG2 phosphor converted white LEDs) have already been successfully applied to optimize the first demonstrator luminaire designs of the Delphi4LED project, [11]. In Figure 9 the calculated results for the SME-version of the first project demonstrator luminaire (corresponding to the so-called "10 W outdoor LED spot" category) are shown. The prototype of the luminaire has also been manufactured with the chosen number of LEDs and with the chosen substrate material and layout arrangement and heatsink. This luminaire prototype with the electrical driver and luminaire optics with the given efficiency (see among the design inputs in Figure 9) was also measured by an independent testing laboratory and good agreement between the measured and simulated major operating parameters was found, see Table 2.

5. Conclusions

In this paper we presented implementation of the "current driven" version of the improved chip level multi-domain LED model proposed earlier for the Delphi4LED project [8], [9]. Both a Visual Basic macro and a generic SPICE subcircuit macro implementation of the model have been prepared and presented. The model has been successfully applied as part of a spreadsheet application to support the design of the first demonstrator luminaire of the Delphi4LED project.

A couple of issues still have to be dealt with, such as the number of LED samples to characterize for extracting the parameter set representative for the given LED type, how to represent the sample-to-sample variations of the model parameters, how to include typical LED product parameters such as binning value of the forward voltage, etc. These are questions that are being still subject of research within the Delphi4LED project [16]. Further issues beyond the scope of the Delphi4LED project are e.g. how to reflect elapsed product life-time in the chip level multi-domain LED model, as proposed by a recent conference paper

	Simulated	Measured
Total input electric power [W]	11.71	10.7
Total emitted luminous flux [lm]	1302	1339

Table 2: Simulated and measured major parameters of the SME-variant of the first Delphi4LED project demonstrator.

Acknowledgments

The work described here has received funding from the European Union's Horizon 2020 research and innovation program through the H2020 ECSEL project Delphi4LED (grant agreement 692465). Co-financing of the Delphi4LED project by the Hungarian government through the NEMZ_16-1-2017-0002 grant of the National Research, Development and Innovation Fund is also acknowledged.

The modelling work of LEDs performed at BME was also supported by the Higher Education Excellence Program of the Ministry of Human Capacities in the frame of Artificial Intelligence research area of Budapest University of Technology and Economics (BME FIKP-MI/SC).

Support from Delphi4LED project partners, especially from G. Martin (Signify), Ch. Marty (Ingelux), D. Fournier (PISEO) and E. Morard (Ecce'lectro) is also acknowledged.

References

[1] The Delphi4LED project website: www.delphi4LED.org

[2] A. Poppe, "*Multi-Domain Compact Modeling of LEDs: an Overview of Models and Experimental Data*", Microelectronics Journal 46(12A): 1138-1151 (2015), DOI: 10.1016/j.mejo.2015.09.013

[3] A. Poppe, "*Simulation of LED Based Luminaires by Using Multi-Domain Compact Models of LEDs and Compact Thermal Models of their Thermal Environment*", Microelectronics Reliability 72(5): 65-74. (2017), DOI: 10.1016/j.microrel.2017.03.039

[4] A. Poppe, J. Hegedűs, A. Szalai, R. Bornoff, J. Dyson, "Creating multi-port thermal network models of LED luminaires for application in system level mul-ti-domain simulation using Spice-like solvers", In: Proceedings of SEMI-THERM'16, 14-17 March 2016, San Jose, USA, pp. 44-49, DOI: 10.1109/SEMI-THERM.2016.7458444

[5] J. Hegedüs, G. Hantos, A. Poppe, "*Light output stabilisation of LED based streetlighting luminaires by adaptive current control*", Microelectronics Reliability, 79(12): 448-456 (2017),
DOI: 10.1016/j.microrel.2017.06.060.

[6] J. Hegedüs, P. Hotváth, T. Szabó, A. Szalai, A. Poppe, "*A New Dimming Control Scheme of LED Streetlighting Luminaires Based on Multi-Domain Simulation models of LEDs in order to Achieve Constant Luminous Flux at Different Ambient Temperatures*", In: Proceedings of the Conference on "Smarter Lighting for Better Life" at the CIE Midterm Meeting 2017., 23-25 October 2017, Jeju, South Korea. Vienna: CIE, 2017. pp. 267-276., (ISBN:978-3-901906-95-4)

[7] J. Hegedüs, P. Horváth, G. Hantos, T. Szabó, A. Szalai, A. Poppe, "*A New Dimming Control Scheme of LED Based Streetlighting Luminaires Using an Embedded LED Model Implemented on an IoT Platform to Achieve Constant Luminous Flux at Different Ambient Temperatures*", In: Matej B Kobav (editor), Proceedings of Lux Europa 2017. Conference, Ljubljana, Slovenia, 18-20 September 2017. Ljubljana: Lighting Engineering Society of Slovenia, 2017. pp. 87-92., (ISBN:978-961-93733-4-7)

[8] G. Hantos, J. Hegedüs, M. C. Bein, L. Gaál, G. Farkas, Z. Sárkány, S. Ress, A. Poppe, M. Rencz, "*Measurement issues in LED characterization for Delphi4LED style combined electrical-optical-thermal LED modeling*", In: Proceedings of EPTC'17, 6-9 December 2017, Singapore, DOI: 10.1109/EPTC.2017.8277493

[9] G. Farkas, L. Gaál, M. Bein, A. Poppe, S. Ress, M. Rencz, "*LED characterization within the Delphi-4LED Project*", In: Proceedings of ITHERM'18, 29 May - 1 June 2018, San Diego, USA, DOI: 10.1109/ITHERM.2018.8419602

[10] P. Raynaud, "*Single Kernel Electro-Thermal IC Simulator*", In: Proceedings of the 19th International Workshop on THERMal INvestigation of ICs and Systems, 25-27 September 2013, Berlin, Germany, pp. 356-358, DOI: 10.1109/THERMINIC.2013.6675231

[11] C. Marty, J. Yu, G. Martin, R. Bornoff, A. Poppe, D. Fournier, E. Morard, "*Design flow for the development of optimized LED luminaires using multi-domain compact model simulations*", In: Proceedings of THERMINIC'18, 26-28 September 2018, Stockholm, Sweden, DOI: 10.1109/THERMINIC.2018.8593318

[12] JEDEC JESD51-51 Standard "Implementation of the Electrical Test Method for the Measurement of Real Thermal Resistance and Impedance of Light-Emitting Diodes with Exposed Cooling" (2012)

[13] G. Hantos, J. Hegedüs, A. Poppe, "*Different questions of today's LED thermal testing procedures*", In: Proceedings of the 34th IEEE Semiconductor Thermal Management Symposium (SEMI-THERM'18), 19-23

March 2018, San Jose, USA, pp. 63-70, DOI: 10.1109/SEMI-THERM.2018.8357354

[14] R. Bornoff, T. Mérelle, J. Sari, A. Di Bucchianico, G. Farkas, "*Quantified Insights into LED Variability*", In: Proceedings of THERMINIC'18, 26-28 September 2018, Stockholm, Sweden,
DOI: 10.1109/THERMINIC.2018.8593315

[15] T. Mérelle, R. Bornoff , G. Onushkin , L. Gaál , G. Farkas, A. Poppe, G. Hantos, J. Sari, A. Di Buc-chianico, "*Modeling and quantifying LED variability*", In: Proceedings of 2018 LED Professional Symposium (LpS2018), 25-27 September 2018, Bregenz, Austria (ISBN: 978-3-9503209-9-2), pp. 194-206

[16] T. Merelle, K. Sari, A. Di Bucchianico, G. Onushkin, R. Bornoff, G. Farkas, L. Gaál, G. Hantos, J. Hegedüs, G. Martin, A. Poppe, "*Does a single LED bin really represent a single LED type?*", submitted to the CIE 2019 29th QUADRENNIAL SESSION, 14-22 June 2019, Washington DC, USA

[17] J. Hegedüs, G. Hantos, A. Poppe, "Lifetime Iso-flux Control of LED based Light Sources Paper", In: Proceedings of the 23rd International Workshop on THERMal INvestigation of ICs and Systems (THERMINIC'17), 27-29 September 2017, Amsterdam, The Netherlands,
DOI: 10.1109/THERMINIC.2017.8233816

Assessment of Critical Heat Flux on Finite Size Surfaces Under Pool Boiling

Julia Reed and Vijay K. Dhir

Department of Mechanical and Aerospace Engineering, University of California, Los Angeles
420 Westwood Plaza, Engineering IV 38-137H
Los Angeles, USA
Jreed473@ucla.edu

Abstract

Boiling heat transfer is an efficient yet not fully understood mode of heat transfer. Many studies have been reported that claim high critical heat fluxes from microstructured surface and new correlations have been developed to explain the enhancement. Data from literature has been compared to isolate the effect from microstructured surfaces. The effect of heater wettability and size is examined. The purpose of this study is to isolate the sources of increased critical heat flux; wettability, heater size, or surface modification. It has been found that the heater size and surface modification effects are inter-mixed.

Keywords

Maximum heat flux, pool boiling, finite surface, microstructure

Nomenclature

- σ   surface tension, N/m
- K   correction factor
- φ   Apparent Contact Angle , degree
- q"   heat flux, W/m²
- ρ   density, kg/m³
- l_c capillary length, m
- β   dynamic receding contact angle, degree
- $λ_d$ Most dangerous wavelength, m
- g   acceleration due to gravity, m/s²
- h_{fg} heat of vaporization, J/kg
- r   roughness factor
- L'   dimensionless length (L'=L/l_c)
- ψ   orientation, degrees

1. Introduction

For both efficiency and safety purposes, increasing critical heat flux is a topic of interest. Nucleate boiling is a very efficient mode of heat transfer which makes it an excellent choice for cooling purposes. Electronics need more and more cooling which leads to the need to achieve higher heat fluxes. [1] Boiling heat transfer is divided into partial nucleate, fully developed nucleate, transitional, and film boiling. The maximum heat flux during nucleate boiling marks the onset of transition boiling. In transition film boiling, the heat flux rapidly decreases as the surface temperature increases, making it both as an unstable and less efficient mode of heat transfer. Film boiling on the other hand is accompanied by high thermal resistance. In order to delay onset of transition boiling, researchers have investigated the use of surface modification with microstructures.

1.1 Models and Correlations

Kutateladze developed an expression for the prediction of maximum heat flux through dimensional analysis and Zuber found nearly the same expression from hydrodynamic considerations which is shown below. [2] Lienhard and Dhir continued with this model using the most dangerous wavelength and extended it to different geometries. [3] Different values were found for the constant C. These theories consider hydrodynamic instability of away from the heater surface as the operating mechanism for critical heat flux in which the vapor removal reaches a critical velocity so that the vapor production rate exceeds that of the vapor removal rate. Vapor removal limit as obtained by Zuber with C=$\frac{\pi}{24}$ is

$$q''_{CHFz} = C * \rho_v h_{fg} \sqrt[4]{\frac{\sigma g (\rho_l - \rho_v)}{\rho_v^2}}$$

While for many years it was widely accepted, currently it is questioned. Dhir and Liaw established a unified approach that encompasses both nucleate and transition boiling based on the void fraction at the wall. [4] Dhir and Liaw [4] discuss how the critical heat flux can either be limited hydrodynamically or by vapor generation. For a surface that is not well wetted the limit is determined by vapor generation.

Kandlikar developed a model for critical heat flux. [5] He used a force balance parallel to the heater on a bubble to predict critical heat flux which occurs when the force due to change in vapor momentum becomes larger than the sum of all other parallel forces causing the bubble to blanket the surface. His model shows the dependence on contact angle and orientation, and was extended to include subcooling. The contact angle used is the dynamic receding contact angle. When conducting his derivation, he used the Taylor critical wavelength as opposed to the most dangerous wavelength. He compared his model to previous models and experimental data to validate his model. The correlation is shown below.

$$q''_{CHF_{Kandlikar}} = K_{Kandlikar} * h_{fg} \rho_g^{\frac{1}{2}} \left(\sigma g (\rho_l - \rho_g) \right)^{\frac{1}{4}}$$

$$K_{Kandlikar} = \left(\frac{1 + cos\beta}{16} \right) \left[\frac{2}{\pi} + \frac{\pi}{4} (1 + cos\beta) cos\psi \right]^{\frac{1}{2}}$$

Wenzel looked at the relationship between roughness and wetting. [6] He found that increasing roughness increases wettability in hydrophilic surfaces and decreases wettability in hydrophobic surfaces. He defined a roughness factor which was the ratio of the actual surface area to the geometric surface area. He discussed the importance of clearly defining roughness. [7]

Recently, many researchers have looked into the effect of micro/nanostructures on nucleate boiling and maximum heat flux. Many have modified the model by Kandlikar to extend to

		q/q_{maxZ}		
Author	Contact Angle	0°	90°	
Kandlikar [5]		1.43	0.27	$K_{Kandlikar} = \left(\dfrac{1+cos\beta}{16}\right)\left[\dfrac{2}{\pi}+\dfrac{\pi}{4}(1+cos\beta)cos\psi\right]^{\frac{1}{2}}$
Chu et al. [8]		1.82	0.43	$K_{Chu} = \left(\dfrac{1+cos\phi}{16}\right)\left[\dfrac{2*(1+\alpha)}{\pi(1+cos\phi)}+\dfrac{\pi}{4}(1+cos\phi)cos\psi\right]^{\frac{1}{2}}$ $\alpha = r\, cos\theta_{receding\ contact\ angle\ (smooth)}.$
Quan et al. [9]		1.97	0.80	$K_{Quan} = \left(\dfrac{1+cos\theta}{16}\right)\left[\dfrac{2}{\pi}\left(1-\sqrt{\phi_s}\right)^{-\frac{1}{2}}\dfrac{r+cos\theta}{1+cos\theta}\right.$ $\left.+\dfrac{\pi}{4}\left(1-\sqrt{\phi_s}\right)^{\frac{1}{2}}(1+cos\theta)cos\psi\right]^{\frac{1}{2}}$
Li and Huang [10]	Micro-pillars	2.34	0.27	$q^{"}_{CHF_{Li}} = q^{"}_{CHF_{Kandlikar}}*S + M*h_{fg}$ $*h\sqrt{\dfrac{P-d}{P^2}g\,cos\theta\,\rho_l\Delta\rho}\dfrac{(P-d)d}{sin\alpha\,P^2}$
	Micro-rough	5.10	0.27	$q^{"}_{CHF_{Li}} = q^{"}_{CHF_{Kandlikar}}*S + M*h_{fg}*h\sqrt{\dfrac{P-d}{P^2}g\,cos\theta\,\rho_l\Delta\rho}$
Ahn et al. [11]		1.18	0.82	$q^{"}_{CHF_{Ahn}} = q^{"}_{CHF_{Kandlikar}}*S$ $+\dfrac{1}{4}\pi(kR)^2\beta^3\left(3\beta\left(\dfrac{dR}{dt}\right)+4kR\left(\dfrac{d\beta}{dt}\right)\right)\dfrac{\rho_l h_{fg}}{A_{heating}}$

Table 1 Correlations for CHF from literature calculated for contact angle 0° and 90°

the effects of these enhanced surfaces. Chu et al. [8] modified Kandlikar's force balance with Wenzel's roughness factor shown in Table 1 and uses the apparent contact angle instead of the dynamic receding contact angle which was common to do among those who modified Kandlikar's equation. Quan et al. [9] included the capillary wicking force. In K_{Quan}, φ_s is the solid fraction. Ahn et al. [11] and Li and Huang [10] choose to add an additional term to Kandlikar's model with pitch, P, diameter, d, height, h, and multiplier, M which was found by fitting to experimental data. Ahn et al. [11] multiply the critical heat flux from Kandlikar's model by constant S which was determined through fitting with their experimental data. They added a liquid spreading term where R is the time averaged radius of curvature and k is an effective parameter to fit the data. The correlations are summarized in Table 1 with the critical heat flux ratio determined through each correlation for a contact angle of 0° and 90°. For Ahn et al. [11], 1° was used to obtain a result. Li and Huang [10] did not give a value for the correction factor to Kandlikars equation, S, so 1 was used. This factor was supposed to correct for the structures.

$$q^{"}_{CHF_{Modified\ Kandlikar}} = K*h_{fg}\rho_g^{\frac{1}{2}}[\sigma_{lv}g*(\rho_l-\rho_g)]^{\frac{1}{4}}$$
Or
$$q^{"}_{CHF_{Modified\ Kandlikar}} = q^{"}_{CHF_{Kandlikar}} + q^{"}_{added}$$

All of the correlations contain the hydrodynamic term from Zuber's model so that they include both surface and hydrodynamic effects.

1.2 Experimental Work

Nanowires, shown in Figure 1 [12], have been investigated. Chen et al. [12] and Lu et al. [13] looked at Si nanowire arrays with water and achieved a maximum critical heat flux of 224 W/cm² on a square heater with L'= 2. The measured a zero contact angle shown in Figure 2 [12]. Li et al. [14] achieved 160 W/cm² with a lower super heat using copper nanorods. They reported a contact angle of 38° and did not report the heater size.

Other geometries of microstructures have been studied with water. Chu et al. [8] achieved a critical heat flux with water of 208 W/cm² on a fully wetted, L' =8, square plate which was claimed to be infinite by the authors. This claim of an infinite plate is consistent with some findings but not all which will be discussed in further detail in section 3.2. The micropillars are shown in Figure 3. [8] Kwak et al. [15] looked at high aspect ratio, rectangular microchannels that were 30 μm wide with varying heights spanning 10 to 100 μm with a contact angle of 35° and achieved a heat flux 250 W/cm² on a 8 x 6, in dimensionless lengths, heater. Li and Huang[10] developed two models based on microstructure geometry, pillars or ridges. Son et al. [16] studied at controlled microroughness as achieved a critical heat flux of 198 W/cm² on a fully wetted 10 x 4, in dimensionless lengths, heater.

Research has been done that looks at nano, micro, and hierarchical structures. Chu et al. [17] continued their research on L'=8 square plates and achieved a critical heat flux of 250

Author	surface type	L'	fluid	q_{max}/q_{maxz}	Reported Contact angle [°]
Chu et al. [8] 2012	micropillar	8 x8	water	1.89	0
	Plain	8 x 8		.68	32
Chu et al. [17] 2013	micropillar w/ nanorods	8 x 8	water	2.15	0
				2.27	0
	Plain	8 x 8		.75	32
Li and Peterson [18] 2010	modulated porous	D =3	water	4.09	N/A
Li, C et al. [14] 2008	nanorods	N/A	water	1.45	38.5
Chen et al. [12] 2009	nanowire	4 x 4	water	1.75	0
		4 x 4		1.79	0
	Plain	4 x 4		.75	40
	Plain	4 x 4		.97	15
Lu, M. C. et al. [13] 2011	nanowire	2 x 2	water	2.04	0
		4 x 4		1.37	0
		6 x 6		1.14	0
		8 x 8		1.14	0
	Plain	2 x 2		.74	40
	Plain	4 x 4		.61	40
	Plain	6 x 6		.42	40
	Plain	8 x		.40	40
Ha and Graham [19] 2017	microporous	8 x 8	water	1.72	N/A
	microporous w/channels	8 x 8		3.23	N/A
Kim, S. et al. [20] 2010	octagonal micro-post	8 x 10	water	1.5	81
	nanorods	8 x 10		1.82	0
	micro-post w/ nanorods	8 x 10		2.12	0
	Plain	8 x 10		.96	83
	Plain	8x 10		1.09	83
Ahn et al. [11] 2011	micro	4x 4	water	1.55	0
	nano	4 x 4		1.36	2-9.2
	micro/nano	4 x 4		1.77	0
	Plain	4 x 4		1.12	23
	Plain	4 x 4		1.01	32
	Plain	4 x 4		0.99	33
	Plain	4 x 4		0.90	49
Kwak et al. [15] 2018	microchannels	6 x 8	water	2.28	N/A
	Plain	6 x 8		1.36	35
Theofanous et al. [21], [22] 2002	rough (not structured)	8 x 16	water	1.55	12-15
Son et al. [16] 2017	controlled microrough	4 x 10	water	1.80	0
Jaikumar and Kandlikar [23] 2015	porous coated microchannels	4 x 4	water	2.85	N/A
Zou and Maroo [24] 2013	Nano/microridge	3 x 5	water	1.64	N/A

Table 2 Experimental results from Literature

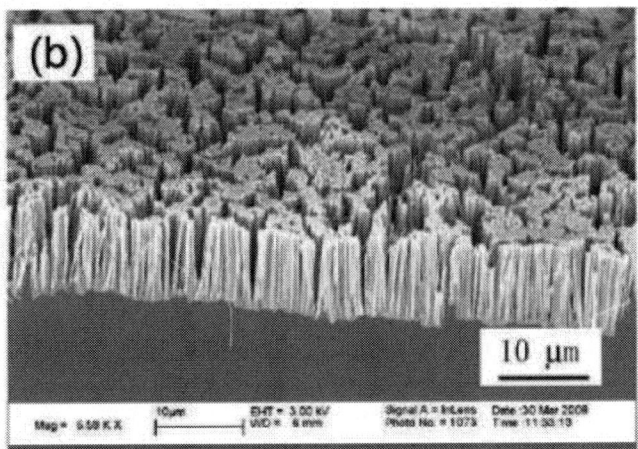

Figure 1 Nanowire from Chen et al. [12]

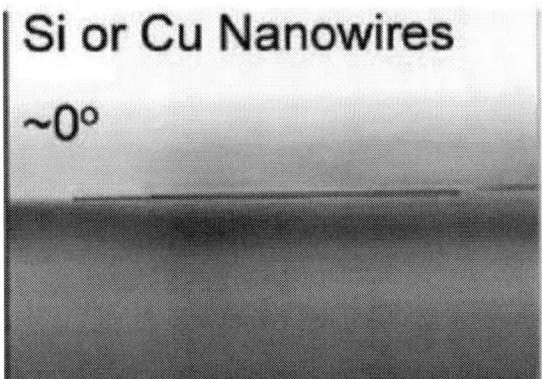

Figure 2 Contact angle from Chen et al. [12]

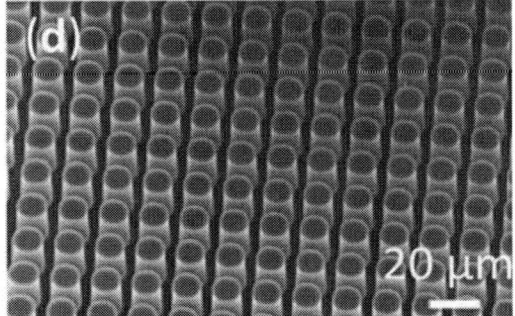

Figure 3 Micropillar from Chu et al.[8]

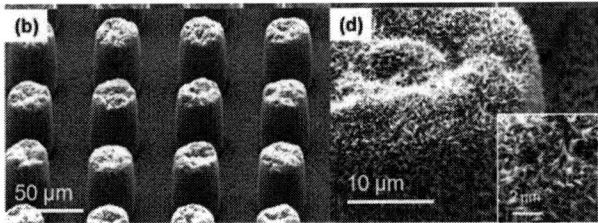

Figure 4 Hierarchical structures from Chu et al. [17]

on L'=8 square plates and achieved a critical heat flux of 250 W/cm² with a hierarchical structure shown in Figure 4. S. Kim et al. [20] achieved a critical heat flux of 233 W/cm² on a dimensionless 10 x 8 heater with a zero contact angle. Because the heater is not square, the length to compare is not as obvious

and the amount that the result exceeds the Zuber limit for finite surfaces of Lienhard et al. [25] depends on which side is look at. Looking shorter length side the result is only at 6% increase while the longest side is almost 70% increase from the finite plate model of Lienhard et al. [25] More detail on size effects and this model will be given in section 2.2. Ahn et al. [11] only achieved 195 W/cm² on fully wetted dimensionless 4 x 4 heater.

Other studies that looked into nanofluids, porous features, and surface condition have been conducted. S. J. Kim et al. [26] looked at the effect of nanofluids and found that the buildup of a porous layer from nanoparticle precipitation as a consequence of nucleate boiling from the evaporation of the microlayer improves the wettability of the surface. Nanoparticle thin film coatings was investigated by Forrest et al. [27] For all of the thin film coatings, the critical heat flux was increased. The boiling curve shifted to the left in the case of the hydrophobic coating. It should be noted that while the contact angle for the hydrophobic coating are 140° static and 160° advancing, the receding contact angle is only 20°. In a set of companion papers, Theofanous et al. [21], [22] described an extensive experiment to better understand the mechanisms involved in critical heat flux and concluded that the critical heat flux is not limited by external hydrodynamics rather it is limited by microhydodynmaics and rupture of the liquid microlayer. Ha and Graham [19], on a microporous structure with channels on a 8L' x 8L' plate that they claim to be infinite, achieved a critical heat flux of 355 W/cm²; no information on wettability was given. Li and Peterson [18] looked at porous structures made of sintered copper and achieved a critical heat flux of 450 W/cm². The heater diameter was L'=3 and no contact angle information was given. Jaikumar and Kandlikar [23] looked at porous coatings on microchannels with channel width 762 µm, channel depth 400 µm and fin width 200 µm on L'=4 heater. They achieved a maximum heat flux of 313 W/cm² and did not report wettability.

The experimental results have been summarized in Table 1. Many of the researchers report a zero contact angle for the structured surface and the experiments where conducted on small heaters.

2.1 Effect of Contact Angle

The effect of wettability must be established in order to determine what enhancements to critical heat flux come from micro/nanostructures or from the enhanced wettability that comes from the structures. The data from Maracy and Winterton [28], Ahn et al. [11], Lu et al. [13], and Chen et al. [12] have been compiled in figure 5 below showing a trend of increasing critical heat flux with increased wettability for a constant size heater. The same lab produced the data points for the 40° contact angle which has dip. The most extensive data is from Maracy and Winterton [28]. The trend can be seen in both the data from Maracy and Winterton [28] and Ahn et al. [11]. The work of Ahn et al. [11] resulted in a trend of higher values for the critical heat flux at the various contact angles than that of Maracy and Winterton [28] suggesting that the size effects play a role on partially wetting surfaces so that the points are shifted up. Many researchers have compared their results for plates with modified surfaces to a plain surface of the same size but with a different contact angle. This highlights the

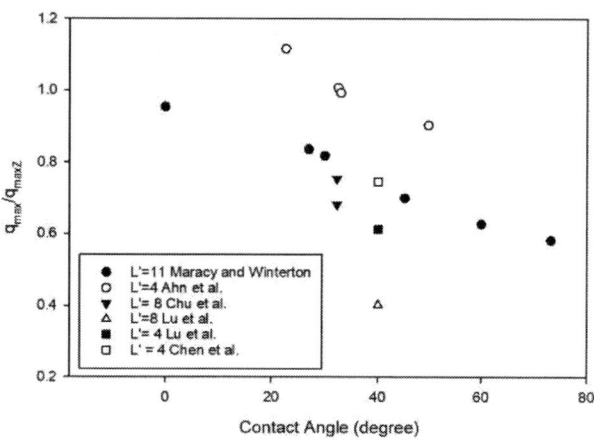

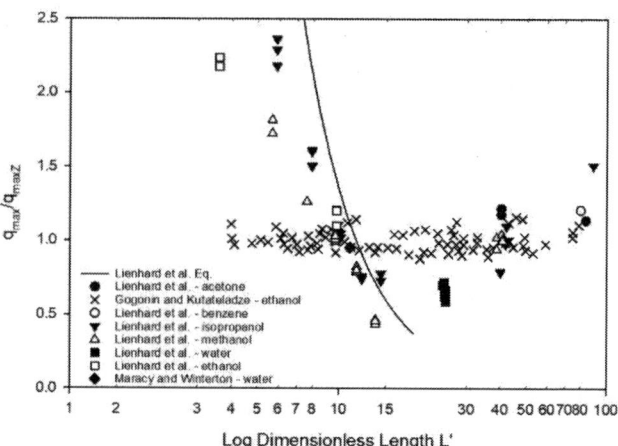

Figure 5 Dimensionless critical heat flux for various contact angles for horizontal plain surfaces from Chu et al. [8], [17] Lu et al. [13] Chen et al.[12] Maracy and Winterton [28] Ahn et al. [11]

importance of comparing the same contact angle when quantifying the enhancement from structured surfaces.

2.2 Effect of Heater Size

There is no consensus on the minimum size to be considered an infinite heater. Lienhard et al. [25] experimentally showed that size effects are present on heaters less than three most dangerous wavelengths, $\lambda_d = 2\pi \left[\frac{3\sigma}{g(\rho_l - \rho_v)}\right]^{\frac{1}{2}}$. They presented a correlation for small heaters greater than one λ_d, where there is no issue of viscous film crowding, [13], [25], and showed that small heaters can greatly exceed the Zuber limit for an infinite plat. This correlation is curve plotted in figure 6.

$$\frac{q_{max}}{q_{maxZ}} = 1.14 \frac{N_j}{\frac{A_H}{\lambda_d^2}}$$

Gogonin and Kutateladze [29] presented data that some interpret as size effects being present wbelow 2 or 3 capillary lengths, $l_c = \left[\frac{\sigma}{g(\rho_l - \rho_v)}\right]^{\frac{1}{2}}$, which is about .2 λ_d. Lu et al. [13] determined that a plate needs to be eight capillary lengths to be considered infinite, but they only looked at small heaters that are less than one λ_d. Numerical simulations suggest that 12 capillary lengths should suffice for an infinite heater below that length would increase the critical heat flux; they simulated partially and poorly wetting surfaces [30], [31]. Figure 6 shows the experimental data for a plain surfaces of different size. Approaching one most dangerous wavelength or approximately 11 capillary lengths, the data of Gogonin and Kutateladze [29], Lienhard et al.[25], and Maracy and Winterton [28] are in agreement showing that the q_{max}/q_{maxz} is about 1. However, as the data moves away from that range, it becomes very contradictory. Gogonin and Kutateladze [29] shows that as long as the heater is more than 4 capillary lengths

Figure 6 Dimensionless critical heat flux . variation with dimensionless heater length for well wetted plain surfaces from Lienhard et al. [25] Gogonin and Kutateladze [29] and Maracy and Winterton [28]

there are no size effects. Meanwhile, Lienhard et al. [25] shows that the size effects are very present on both sides of the range. To the left the q_{max}/q_{maxz} increases and more than does and to the right the critical heat flux decreases then returns to about 1. The data somewhat levels after 11 capillary lengths. The data of Lienhard et al. [25] has more variation in the region of not size effect >32 l_c. Theofanous et al. [21] took images that showed uniform boiling and added a crate-like structure that divided the plate into 8 square with L'=4 cells to show no difference from the undivided heater surface in order to confirm Gogonin and Kutateladze. [29] Based on this data, no clear answer can be seen for where size effects set in.

The data available from literature on structured surfaces are all less than 11 capillary lengths which is the region of discrepancy in data for the size effects because while Lienhard et al. [25] clearly showed size effects in this region, Gogonin and Kutateladze [29] clearly showed that there would be no size effects through 4 capillary lengths. Lu et al. [13] investigated the size effects but did not extend past this range. Figure 7 has

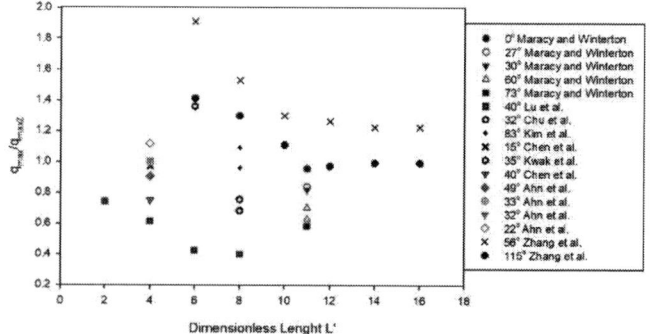

Figure 7 Size effects for plain surfaces with various contact angles. The data of Zhang et al. [30] was from a numerical simulation. Experimental data is from Maracy and Winterton [28], Lu et al. [13], Chu et al. [8], [17], Kim et al. [20], Chen et al. [12], Kwak et al. [15], and Ahn et al. [11].

22

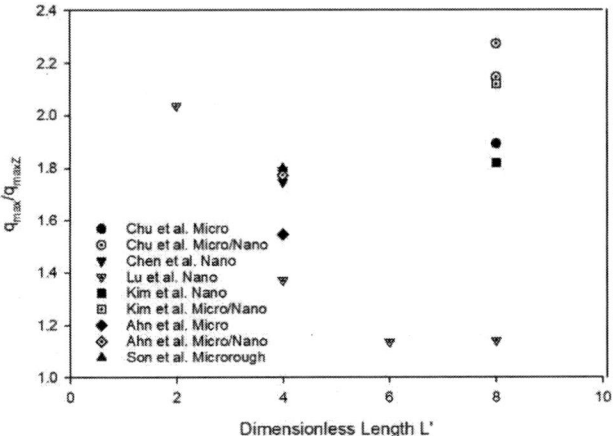

Figure 8 Microstructured Surfaces with reported contact angles of zero with water as the working fluid. The data comes from Son et al. [16], Ahn et al. [11], Chen et al. [12], Chu et al. [8], Chu et al. [17], and Kim et al. [20].

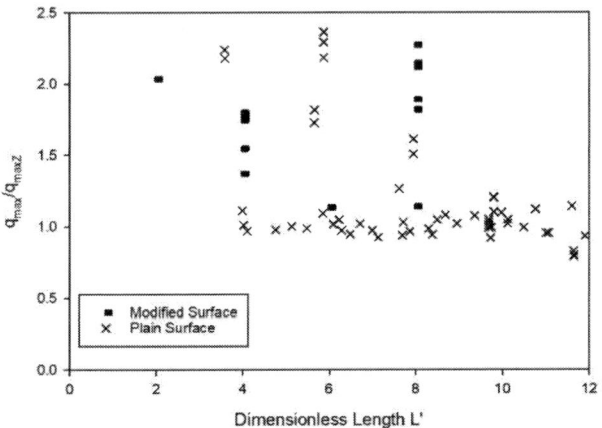

Figure 9 All the modified surface data has a reported contact angle of zero. The fluid is water unless otherwise indicated. The work of Lienhard et al. [25] and Gogonin and Kutateladze [29] was on plain surfaces. The modified surface data comes from Son et al. [16], Ahn et al. [11], Chen et al. [12], Chu et al. [8], Chu et al. [17], and Kim et al. [20].

effects through 4 capillary lengths. Lu et al. [13] investigated the size effects but did not extend past this range. Figure 7 has size effects plotted for various contact angles. The trend of shifted size effects curves can be seen for 40° and most of the points for 30±5°. The numerical points from Zhang et al. [30], [31] are higher than would be expected given the contact angles.

2.3 Effect of Micro/Nanostructured Surfaces
To see the effects of micro/nano-structures, the data must be looked at for a constant heater size and constant contact angle, shown in Figure 9, so enhancements to critical heat flux are not falsely attributed to structure that are simply due to

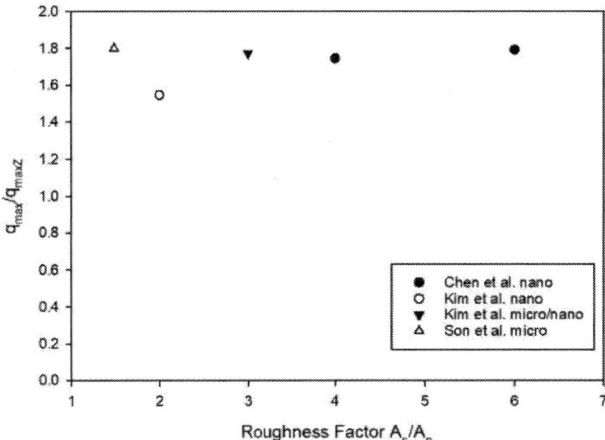

Figure 10 The relationship between roughness factor and critical heat flux for fully wetted surfaces 4 capillary lengths wide from Chen et al. [12],, Kim et al. [20], and Son et al. [16]

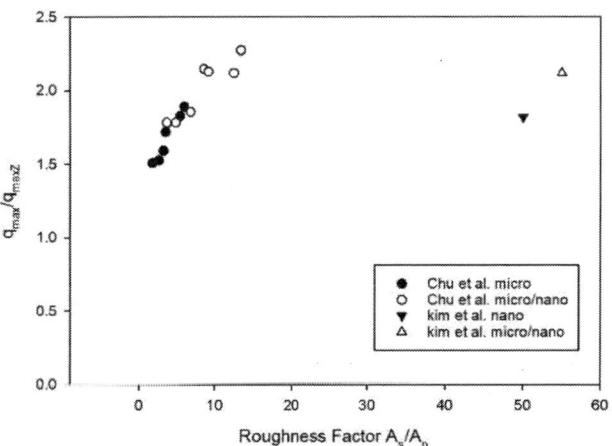

Figure 11 The relationship between roughness factor and critical heat flux for fully wetted surfaces 8 capillary lengths wide from Chu et al. [8], [17], and Kim et al. [20].

increased wettability or heater size effects. A close look was taken to see if the structures gave enhancement beyond what would have been expected for a fully wetting plain surface. Due to the discrepancy in the heater size data on a plain surface, none of the data can be treated as an infinite plate. To isolate the effects of microstructures, the heaters need to be treated as finite by looking each dimensionless length individually. Figure 8 shows the data for only microstructured surfaces. Most of the data for microstructured surfaces is at 4 and 8 capillary lengths so these are the points that were most closely examined. At 4 capillary lengths, the modified surface data fits within the existing data for small, well wetting, plain surface heaters and is indistinguishable from that of the plain surface. This contradicts the claim of Lu et al. [13] that the size and structure effects are simultaneous and additive. From the conclusions of Lu et al. [13], one would expect to see all the microstructured data to lay above any plain surface data in Figure 9. Figure 11 shows the roughness factor show no trend to change critical heat flux at 4 capillary lengths which is contradictory to the claims on Chu et al. [8] who expanded Kandlikar's model to include roughness factor. At 8 capillary lengths, all but one modified surface has a higher critical heat flux than that of the plain surface (Figure 9). There is variation in improvement from the structure which were all reported to be fully wetting. This is indicative of structured surfaces increasing in critical heat flux as it is higher than that of a well wetting plain surface, but this cannot be extended to any point or used as a prediction. For the 8 capillary length surface, for lower values of roughness factor it appears that critical heat flux increases with roughness factor, but the roughness factor has a large data gap (Figure 10). The data shows a lot of scatter. Because the effects are not consistent for the different heater sizes no predictions can be made on how modifying a surface would alter the critical heat flux at any given heater size.

3. Conclusions

While critical heat flux is important in heat transfer, the mechanisms involved are still not fully understood. It can be stated that the increase wettability increases critical heat flux for partially wetted surfaces. The point at which heater size effects begin is not known due to contradictory reports. The role of microstructures is not known. In order to properly understand the role of microstructures, their effects must be separated from heater size. There is no data for microstructured surface that can be considered by all studies as infinite. While investigating microstructures, many researchers over-report the enhancement due to modified surfaces by comparing the critical heat flux to a plain surface without fixing the other parameters such as size and contract angle. For future studies, more care is required to isolate the effects of structured surfaces. This can be done by fully characterizing how the heater surface affects critical heat flux for its size and response to wettability. After looking at the fully wetted surface, structures can then be looked at, facilitating a clear comparison with a surface in which other influencing factors can be eliminated.

References

[1]   J. R. Thome, "The new frontier in heat transfer: Microscale and nanoscale technologies," *Heat Transfer Engineering*, vol. 27, no. 9, pp. 1–3, 2006.

[2]   V. K. Dhir, "BOILING HEAT TRANSFER," *Annu. Rev. Fluid Mech.*, vol. 30, no. 1, pp. 365–401, 1998.

[3]   J. H. Lienhard and V. K. Dhir, "Hydrodynamic Prediction of Peak Pool-boiling Heat Fluxes from Finite Bodies," *J. Heat Transfer*, vol. 95, no. 2, p. 152, 1973.

[4]   V. K. Dhir and S. P. Liaw, "Framework for a Unified Model for Nucleate and Transition Pool Boiling," *J. Heat Transfer*, vol. 111, no. 3, p. 739, 1989.

[5]   S. G. Kandlikar, "A Theoretical Model to Predict Pool Boiling CHF Incorporating Effects of Contact Angle and Orientation," *J. Heat Transfer*, vol. 123, no. 6, pp. 1071–1079, 2001.

[6]   R. N. Wenzel, "Resistance of solid surfaces to wetting by water," *Ind. Eng. Chem.*, vol. 28, no. 8, pp. 988–994, 1936.

[7]   R. N. Wenzel, "Surface Roughness and Contact Angle.," *J. Phys. Colloid Chem.*, vol. 53, no. 9, pp. 1466–1467, 1949.

[8]   K. H. Chu, R. Enright, and E. N. Wang, "Structured surfaces for enhanced pool boiling heat transfer," *Appl. Phys. Lett.*, vol. 100, no. 24, p. 241603, 2012.

[9]   X. Quan, L. Dong, and P. Cheng, "A CHF model for saturated pool boiling on a heated surface with micro/nano-scale structures," *Int. J. Heat Mass Transf.*, vol. 76, pp. 452–458, 2014.

[10]   R. Li and Z. Huang, "A new CHF model for enhanced pool boiling heat transfer on surfaces with micro-scale roughness," *Int. J. Heat Mass Transf.*, vol. 109, pp. 1084–1093, 2017.

[11]   H. S. Ahn, H. J. Jo, S. H. Kang, and M. H. Kim, "Effect of liquid spreading due to nano/microstructures on the critical heat flux during pool boiling," *Appl. Phys. Lett.*, vol. 98, no. 7, p. 071908, 2011.

[12]   R. Chen, M.-C. Lu, V. Srinivasan, Z. Wang, H. H. Cho, and A. Majumdar, "Nanowires for Enhanced Boiling Heat Transfer," *Nano Lett.*, vol. 9, no. 2, pp. 548–553, 2009.

[13]   M. C. Lu, R. Chen, V. Srinivasan, V. P. Carey, and A. Majumdar, "Critical heat flux of pool boiling on Si nanowire array-coated surfaces," *Int. J. Heat Mass Transf.*, vol. 54, no. 25–26, pp. 5359–5367, 2011.

[14]   C. Li, Z. Wang, P. I. Wang, Y. Peles, N. Koratkar, and G. P. Peterson, "Nanostructured copper interfaces for enhanced boiling," *Small*, vol. 4, no. 8, pp. 1084–1088, 2008.

[15]   H. J. Kwak, J. H. Kim, B. S. Myung, M. H. Kim, and D. E. Kim, "Behavior of pool boiling heat transfer and critical heat flux on high aspect-ratio microchannels," *Int. J. Therm. Sci.*, vol. 125, no. July 2017, pp. 111–120, 2018.

[16]   H. H. Son, G. H. Seo, U. Jeong, D. Y. Shin, and S. J. Kim, "Capillary wicking effect of a Cr-sputtered superhydrophilic surface on enhancement of pool boiling critical heat flux," *Int. J. Heat Mass Transf.*, vol. 113, pp. 115–128, 2017.

[17]   K. H. Chu, Y. S. Joung, R. Enright, C. R. Buie, and E. N. Wang, "Hierarchically structured surfaces for boiling critical heat flux enhancement," *Appl. Phys. Lett.*, vol. 102, no. 15, p. 151602, 2013.

[18]   C. H. Li and G. P. Peterson, "EXPERIMENTAL STUDY OF ENHANCED NUCLEATE BOILING HEAT TRANSFER ON UNIFORM AND MODULATED POROUS STRUCTURES," *Front. Heat Mass Transf.*, vol. 1, no. 2, 2010.

[19]   M. Ha and S. Graham, "Pool boiling characteristics and critical heat flux mechanisms of microporous surfaces and enhancement through structural modification," *Appl. Phys. Lett.*, vol. 111, no. 9, p. 091601, 2017.

[20]   S. Kim *et al.*, "Effects of nano-fluid and surfaces with nano structure on the increase of CHF," *Exp. Therm. Fluid Sci.*, vol. 34, no. 4, pp. 487–495, 2010.

[21]　T. G. Theofanous, J. P. Tu, A. T. Dinh, and T. N. Dinh, "The boiling crisis phenomenon part I: Nucleation and nucleate boiling heat transfer," *Exp. Therm. Fluid Sci.*, vol. 26, no. 6–7, pp. 775–792, 2002.

[22]　T. G. Theofanous, T. N. Dinh, J. P. Tu, and A. T. Dinh, "The boiling crisis phenomenon part II: Dryout dynamics and burnout," *Exp. Therm. Fluid Sci.*, vol. 26, no. 6–7, pp. 793–810, 2002.

[23]　A. Jaikumar and S. G. Kandlikar, "Enhanced pool boiling heat transfer mechanisms for selectively sintered open microchannels," *Int. J. Heat Mass Transf.*, vol. 88, pp. 652–661, 2015.

[24]　A. Zou and S. C. Maroo, "Critical height of micro/nano structures for pool boiling heat transfer enhancement," *Appl. Phys. Lett.*, vol. 103, no. 22, p. 221602, 2013.

[25]　J. H. Lienhard, V. K. Dhir, and D. M. Riherd, "Peak Pool Boiling Heat-Flux Measurements on Finite Horizontal Flat Plates," *J. Heat Transfer*, vol. 95, no. 4, p. 477, Nov. 1973.

[26]　S. J. Kim, I. C. Bang, J. Buongiorno, and L. W. Hu, "Surface wettability change during pool boiling of nanofluids and its effect on critical heat flux," *Int. J. Heat Mass Transf.*, vol. 50, no. 19–20, pp. 4105–4116, 2007.

[27]　E. Forrest, E. Williamson, J. Buongiorno, L. W. Hu, M. Rubner, and R. Cohen, "Augmentation of nucleate boiling heat transfer and critical heat flux using nanoparticle thin-film coatings," *Int. J. Heat Mass Transf.*, vol. 53, no. 1–3, pp. 58–67, 2010.

[28]　M. Maracy and R. H. S. Winterton, "Hysteresis and contact angle effects in transition pool boiling of water," 1988.

[29]　I. I. Gogonin and S. S. Kutateladze, "Critical heat flux as a function of heater size for a liquid boiling in a large enclosure," *J. Eng. Phys.*, vol. 33, no. 5, pp. 1286–1289, 1977.

[30]　C. Zhang, P. Cheng, and F. Hong, "Mesoscale simulation of heater size and subcooling effects on pool boiling under controlled wall heat flux conditions," *Int. J. Heat Mass Transf.*, vol. 101, pp. 1331–1342, 2016.

[31]　P. Cheng, C. Zhang, and S. Gong, "Lattice Boltzmann Simulations of Macro/Microscale Effects on Saturated Pool Boiling Curves for Heated Horizontal Surfaces," *J. Heat Transfer*, vol. 139, no. 11, p. 110801, 2017.

Molecular Dynamic Simulation of Evaporative Heat Transfer on Graphene Coated Silicon Substrate for Electronics Cooling

Binjian Ma[1], Shan Li[1], Damena Agonafer[1], Baris Dogruoz[2]
1. Washington University in St. Louis, 1 Brookings Dr., St. Louis, MO, USA
2. Cisco Systems Inc., 425 E Tasman Dr., San Jose, CA, USA
PH: 979.492.8888, email: mabinjian@email.wustl.edu

Extended Abstract

SUMMARY

Two-phase cooling such as thin film evaporation is becoming increasingly popular for thermal management of high powered electronics due to the high latent heat associated with the phase change process. Nanoengineered surfaces have been shown to improve two-phase heat transfer performance through enhanced wettability and reduced interfacial thermal resistance. However, how interfacial resistance varies with surface wettability and how such resistance can affect thin-film evaporative transport is still not well understood. In this study, we investigate the evaporative transport characteristics and wetting state of an evaporating thin liquid film on both smooth and nanocoated surfaces using Molecular Dynamics (MD) simulations. The surface wettability between liquid argon and silicon (100) surface coated with 0, 1, and 3 layers of graphene is characterized using equilibrium molecular dynamics methods. The associated interfacial thermal resistances and the evaporation rates are explored using non-equilibrium molecular dynamics methods, in which a hot and cold solid substrate are implemented to facilitate the evaporation and condensation of liquid argon molecules.

Contributions provided by this paper:

1) Applying one or three layers of graphene coatings on a silicon substrate yields 80.7% and 237% higher interfacial thermal resistance between the solid substrate and liquid argon.
2) Addition of one and three layers of graphene reduces the evaporation rate by 38% and 62% times with the same temperature gradient between hot and cold source compared to evaporation form bare silicon surface.
3) Addition of one and three layers of graphene results in an increase in the apparent contact angle from 7° to 13° and 17°, respectively
4) There exists a strong dependence of evaporative mass transport rate on the thermal resistance across the solid-liquid interface and the surface wettability. An increasing level of non-wetting lead to increased interfacial thermal resistance and therefore a lower evaporation rate.

1. INTRODUCTION

Traditional heat dissipation methods based on single phase air or liquid cooling are reaching their limits for keeping the functional unit at a sufficiently low temperature with an ultra-high heat flux (> 1 kW/cm^2) [1-2]. Two-phase cooling such as thin film evaporation, owing to the large amount of latent heat in the phase change process, can effectively remove large amounts of heat while maintaining a small temperature difference across the heat transport system. As the evaporating liquid becomes sufficient thin, the surface characteristics such as interfacial thermal resistance become important parameters affecting the heat and mass transport during thin-film evaporation. This resistance is induced by the acoustic mismatch between the solid and liquid molecules, which impedes for heat propagating across the interface [3-4]. Besides, based on the fundamental theory behind interfacial thermal resistance, the heat exchange across the solid-liquid interface is a strong function of the affinity between the two phases, i.e., the surface wettability. In general, it is expected that interfacial thermal resistance is positively related to the contact angle of the working fluid on the solid substrate. However, it remains unclear if the surface wettability has a direct impact on evaporative transport behavior from a thin liquid film. More importantly, we still lack a general description of the change in interfacial resistance with surface wettability and how such resistance can affect the thin-film evaporative transport. Understanding the relationship between interfacial thermal resistance, surface wettability, and evaporation behavior is important for evaluating the evaporative transport rate on different surfaces and designing rational nanocoatings to enhance evaporative heat transfer.

This study uses equilibrium and non-equilibrium molecular dynamics simulations to comprehensively analyze the surface wettability, interfacial thermal resistance, and the evaporation behavior of a thin film of liquid argon resting on graphene-coated silicon surfaces. The surface wettability is explored by equilibrating a liquid droplet on the substrate at constant temperature condition, while the thermal transport characteristics are analyzed by facilitating the evaporation process with a constant temperature gradient in different liquid-solid combinations and by exploring the dependence of evaporative transport on the surface characteristics.

2. SIMULATION METHOD AND SETUP

The configuration of the evaporation simulation domain is shown in Figure 1(a). Two solid substrates are placed at the top and bottom of the cubic simulation box. Each solid substrate is further covered with 0, 1, and 3 layers of graphene coating and a thin liquid argon layer. The initial interlayer distance between the silicon and graphene was set as 3 Å to avoid any overwhelming intermolecular forces during the early stage of the simulation [5]. Figure 1 shows the top view of the graphene-silicon structure used in our simulation. For all simulation cases, a periodic boundary condition was imposed on both the x and y directions, while a fixed solid wall boundary condition was imposed in the z direction. To maintain a constant simulation volume and prevent the solid substrate from drifting during the simulation, the outermost layer of solid atoms was kept fixed in its coordinates. The subsequent two layers of solid atoms were set as the heat source and heat sink in the bottom and top blocks, respectively, for inducing a temperature gradient to facilitate evaporation. Specifically, the average temperatures of the heat source and heat sink region were kept at two constant values during the evaporation process by rescaling the velocities of the molecules bounded in each region to the designated value after each iteration. The rest of the solid atoms were allowed to move and vibrate freely as a heat conduction medium.

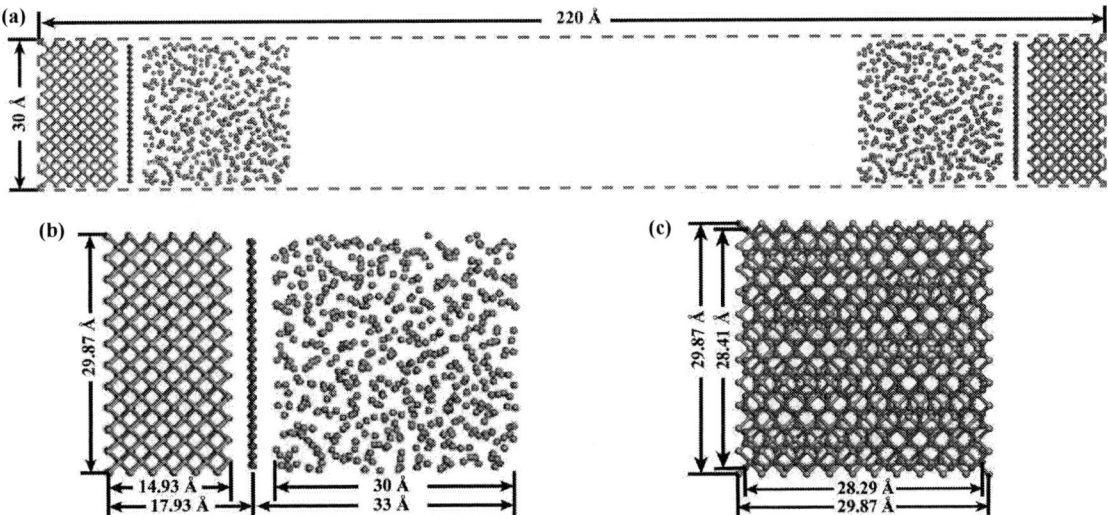

Fig. 1. (a) The initial configuration of the non-equilibrium evaporation using silicon substrate (100) coated with one layer of graphene as an example; (b) the actual dimension of and the spacing between the silicon block, graphene layer, and liquid argon block; and (c) the top view of graphene layer overlapping with the silicon 100 surface.

During the simulation, the non-bonded intermolecular interactions between different atoms were modeled using the standard Lennard-Jones (LJ) potential[6] with a cutoff distance of 12 Å while the bonded interactions are modeled by CVFF forcefield[7]. A local energy minimization step was first performed on the simulation system to mitigate any unphysical initial molecular. Subsequently, the system was equilibrated at an initial temperature of 85 K via a Nosé–Hoover thermostat (NVT ensemble) for 50 ps. Afterwards, the liquid molecules were released from the NVT ensemble and relaxed in an NVE ensemble for 100 ps. Meanwhile, the solid substrates were still held in the NVT ensemble for maintaining a constant temperature. Lastly, the temperature of the heat source and heat sink are rescaled to 100 K and 85 K after each iteration to facilitate the evaporation process. During the evaporation process, the whole simulation domain (excluding the outermost layer of solid atoms) was kept in an NVE ensemble to enable heat propagation from the heat source to the heat sink.

3. RESULTS

Figure 3 shows snapshots of liquid argon evaporating from the heat source (bottom substrate) and condensing on the cold substrate (top substrate) for bare silicon (Figure 3a), silicon coated with one layer of graphene (Figure 3b), and silicon coated with three layers of graphene (Figure 3c). The brown, black, and blue molecules represent silicon, graphene, and argon, respectively. We found that the evaporation rates on silicon, silicon coated with one layer of graphene, and silicon coated with three layers of graphene are 27.48, 17.04, and 10.44 g/(cm²s), respectively. Furthermore, we evaluated the interfacial thermal resistance based on the temperature jump at the solid-liquid interface and the corresponding heat fluxes. Figure 6 shows the temperature profile along the z-direction which exhibits very large fluctuations. We approximated the temperature drop at interface as the difference between the average temperatures inside the solid and liquid domains. The final results are summarized in Table 1. The interfacial thermal resistance between argon and silicon is increases by 80.7% and 237% with one and three layers of graphene coating.

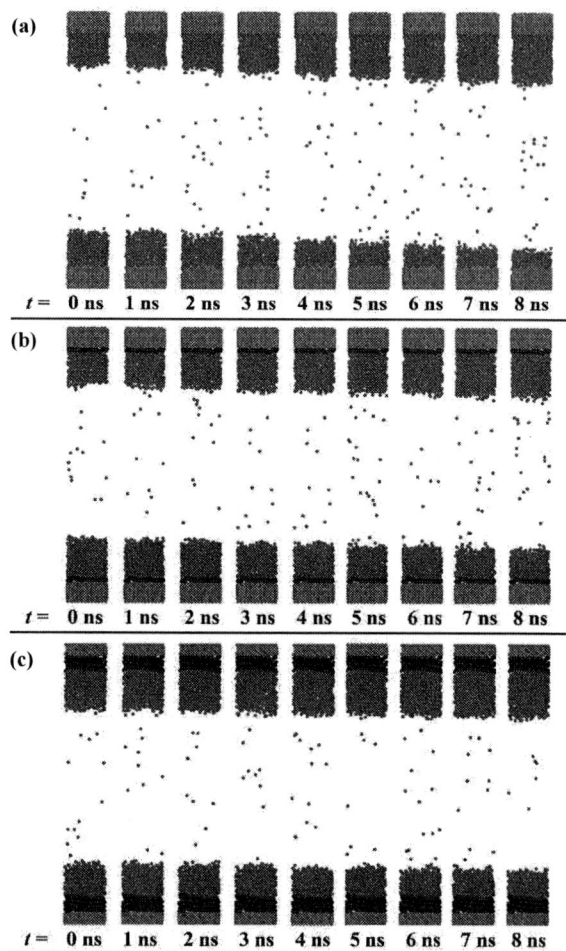

Fig. 2. Snapshots of MD simulation for evaporation of argon on: (a) silicon; (b) silicon coated with one layer of graphene; (c) silicon coated with three layer of graphene.

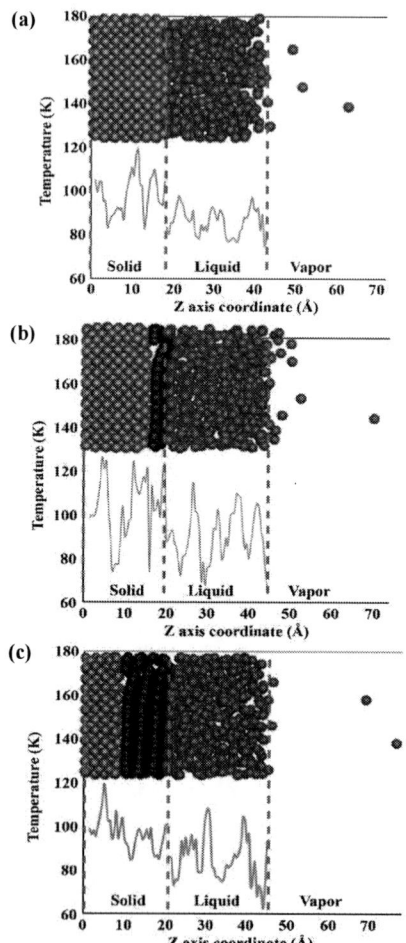

Fig. 3. Temperature profile near the heat source taken along the z-axis for evaporation on (a) silicon; (b) silicon coated with one layer of graphene; (c) silicon coated with three layers of graphene.

Table 1. Evaporative heat flux, interfacial temperature drop, and interfacial thermal resistances for evaporation on different solid surfaces.

	Silicon	Silicon + 1 layer of graphene	Silicon + 3 layers of graphene
q''_{evap} (W/m^2)	0.576	0.382	0.263
ΔT (K)	12.4	13.9	15.9
R_{int} (Km^2/W)	2.15×10^{-7}	3.89×10^{-7}	7.27×10^{-7}

Figure 4 shows the equilibrium shape of the argon droplets resting on silicon surface coated with 0, 1, and 3 layers of graphene. These shapes are obtained by relaxing a liquid argon cube with a size of 50 Å × 50 Å × 50 Å on the substrates at 85 K in an NVE ensemble. As shown in Figure4(b)-(c), the addition of one and three layers of graphene results in an increase in the apparent contact angle from 7° to 13° and 17°, respectively.

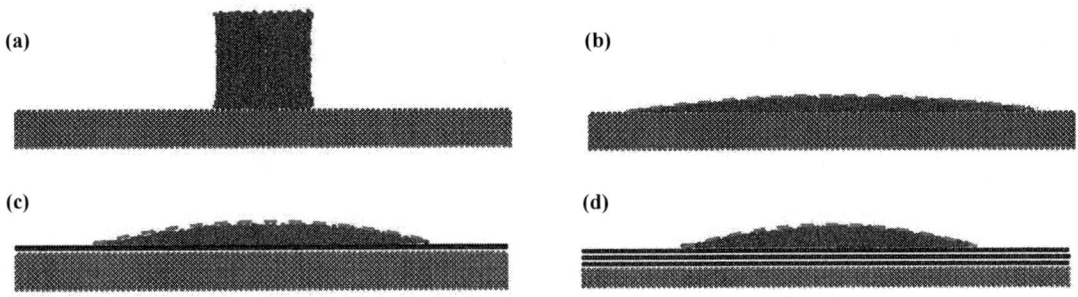

Fig. 4. (a) Initial configuration of liquid argon resting on the substrate; (b)~(d) equilibrium shape of liquid argon on on silicon surface coated with 0, 1, and 3 layers of graphene.

4. CONCLUSIONS

In this study, the evaporative transport behaviors of a thin liquid film from bare silicon, silicon coated with one layer of graphene, and silicon coated with three layers of graphene were explored using molecular dynamics simulation. The results demonstrated a strong dependence of evaporative mass transport rate on the thermal resistance across the solid-liquid interface which was also reflected in an increase in surface hydrophobicity. In particular, addition of one and three layers of graphene on silicon substrate resulted in 38% and 62% reduction in interfacial conductance (or 61% and 163% increases in interfacial resistance) and evaporative mass transfer rate. These studies will provide guidelines to the rational design for surface functionalization in enhancing two-phase heat transfer for a diverse range of engineering and scientific applications.

5. REFERENCES

1. Tuckerman, D. B.; Pease, R. F. W., High-performance heat sinking for VLSI. *IEEE Electron device letters* **1981,** *2* (5), 126-129.
2. Kandlikar, S. G.; Colin, S.; Peles, Y.; Garimella, S.; Pease, R. F.; Brandner, J. J.; Tuckerman, D. B., Heat transfer in microchannels—2012 status and research needs. *Journal of Heat Transfer* **2013,** *135* (9), 091001.
3. Khalatnikov, I., Heat exchange between a solid and He II. *Zhur. Eksptl'. i Teoret. Fiz.* **1952,** *22*.
4. Mahan, G., Kapitza Resistance at a Solid-Fluid Interface. *Nanoscale and Microscale Thermophysical Engineering* **2008,** *12* (4), 294-310.
5. Ramos-Alvarado, B.; Kumar, S.; Peterson, G., On the wettability transparency of graphene-coated silicon surfaces. *The Journal of chemical physics* **2016,** *144* (1), 014701.
6. Lennard-Jones, J. E., Cohesion. *Proceedings of the Physical Society* **1931,** *43* (5), 461.
7. CVFF forcefield file in new format, converted from original format file shipped with Discover 2.6.0/ InsightII 1.1.0/ Insight 2.6. Biosym Technologies, I., Ed. 1990.

Thermal Performance of Metal Foam Heat Sink with Pin Fins for Non-Uniform Heat Flux Electronics Cooling

Yongtong Li[1,2], Liang Gong[*1], Minghai Xu[1], Yogendra Joshi[*2]

1 College of Pipeline & Civil Engineering, China University of Petroleum (East China), Qingdao, China, 266580
2 The George W. Woodruff School of Mechanical Engineering, Georgia Institute of Technology, Atlanta, GA, USA , 30332-0405
*Corresponding authors: lgong@upc.edu.cn; yogendra.joshi@me.gatech.edu

Abstract

In this paper, a concept of metal foam heat sink with pin fins (MFPF heat sink) is presented to improve the cooling performance of high powered electronics with non-uniform heat flux. Numerical simulation is carried out to investigate the MFPF heat sink on the thermal and hydraulic performances in comparison to metal foam heat sink, and traditional pin fin heat sink. The thermal effectiveness of MFPF heat sink is demonstrated by comparing its temperature control capability with the baseline configurations. Based on this, the MFPF heat sink is employed to remove the non-uniform heat flux to ensure the electronic devices operation below a specified maximum temperature. MFPF heat sink under several different power levels are also examined. Results show that MFPF heat sink effectively improves the thermal performance and make the bottom wall temperature more uniform. The thermal performance of metal foam heat sink is more sensitive to porosity than pore density. A local heat flux of 100 W/cm^2 is successfully dissipated using the proposed MFPF heat sink with the junction temperature below 90 °C.

Keywords

Metal foam heat sink, heat transfer, numerical simulation

Nomenclature

Optional listing of terms and units

1. Introduction

Effective cooling of electronic devices is critical to ensure the functionality, improve the performance, and prolong the lifespan of the product. With the emerging trend towards heterogeneous integration, multiple heat dissipating devices with vastly different heat fluxes and acceptable temperature limits are being placed on a common substrate, an interposer, thereby making thermal management a key challenge. For such systems, "site specific" thermal management will be a key driver. Various passive and active cooling techniques have been developed to facilitate heat transfer, such as decreasing boundary layer thickness, intensifying fluid mixing, and increasing heat transfer area. Among these techniques available, increasing internal heat transfer area is the most commonly used method. Several techniques of extending the surface area have been reported by using plate-fins, pin fins, ribs, and porous media [1-4].

As a commonly used porous medium, high porosity metal foam can significantly intensify the fluid mixing, improve the effective thermal conductivity and enhance the convective surface area, and it has great potential as compact heat sink for electronics cooling. Thermal performance of metal foam filled heat sink has been extensively studied for thermal management. Rachedi et al. [5] numerically investigated the cooling performance using foam material, and reported that a temperature reduction of 50% was obtained with foam insert. Bayomy et al. [6] demonstrated the feasibility for cooling high performance Intel i7 processor using water cooled aluminum foam heat sink. Singh et al. [7] demonstrated that a heat flux of 2.9 MW/m2 was successfully removed using a porous sintered heat sink, with a maximum temperature increase of 100 ± 5 °C. Wan et al. [8] observed that the heat flux up to 140 W/cm2 was removed by a novel miniature water-cooled porous heat sink with the heater junction temperature below 62.9 °C. Huang et al. [9-10] employed the porous micro-channel heat sink to cool the high powered electronics and performed numerically investigation of its thermal performance. Results showed that the porous micro-channel heat sink was more effective than the micro-channel heat sink without porous medium under large pumping power condition, and the total thermal resistance was significantly reduced. Boomsma et al. [11], Li et al. [12], and Jiang et al. [13] observed that the forced convection heat transfer in the channel with insertion of metal foam performed up to several times that without foam. Normally, the heat transfer performance of heat sink with metal foam is substantially improved. However, the corresponding high pressure drop is accompanied as a penalty, which restricts the application of metal foam with excellent thermal performance for most engineering cases requiring low pressure loss.

To deal with this issue, the metal foam partially filled heat sink is utilized to reduce the pressure drop though the heat sink without greatly deteriorating the heat transfer. Hadim et al. [14] studied the forced convection in fully and partially filled porous channels with discrete heat sources on the bottom wall. Results demonstrated that when the width of the heat source and the space between the porous layers were of same dimension as the channel height, the heat transfer enhancement in the partially filled channel was almost the same as the fully filled channel, whereas the pressure drop was much lower. Yang et al. [15] numerically analyzed the forced convection in the porous fin heat sink with different fin shapes, and concluded that the heat transfer efficiency of long elliptic porous fin was higher than others. Sener et al. [16] experimentally examined the effects of metal foam filling ratios on heat transfer and pressure drop, with 10 PPI and 20 PPI aluminum foam inserts. They noted that the thermal enhancement factor (TEF) was higher for 10 PPI when the channel was partially filled with aluminum foam. Farsad et al. [17] presented numerical simulations for a micro-channel heat sink filled partially with copper foam.

Apart from filling the channel with metal foam for the improvement of heat transfer, another method of producing a metal foam heat sink is to combine metal foam and solid base material (plate fin, pin fin, or baffle) together [18-22]. Bhattacharya et al. [18] proposed a finned aluminum foam heat sink for electronics cooling, and concluded that significant heat transfer enhancement was obtained by incorporating fins into aluminum foam. Based on the finned foam heat sink proposed by Bhattacharya et al. [18], DeGroot et al. [19] conducted numerical investigation of the forced convective heat transfer in the finned metal foam heat sink, and reported that the finned metal foam heat sink thermally outperformed the plate fin heat sink due to the increased effective thermal conductivity. Seyf et al. [20] numerically analyzed the conjugate heat transfer in aluminum foam heat sink with elliptic pin fins. A heat transfer rate increase up to 400% was obtained compared to that without pin fins. In addition, under forced impinging air jet condition, Feng et al. [21] and Jeng et al. [22] reported that metal foam heat sink with plate fins and copper conductive cylinder could further enhance the heat transfer performance.

As summarized in the above literature review, metal foam heat sink incorporated into the channel fully or partially filled with metal foam could greatly enhance heat transfer. Furthermore, a co-design of heat sink filled with metal foam, i.e., metal foam heat sink with pin fins, can further improve the thermal performance due to the increased heat conduction capability. However, the concept of metal foam heat sink with pin fins partially placed into the channel for handling spatially non-uniform heat fluxeshas not been reported. In this paper, the forced convection heat transfer due to water flow under laminar flow condition is numerically investigated by employing the Brinkman-Forchheimer extended Darcy momentum equation, and the two-model energy equation. Thermal effectiveness of metal foam heat sink with pin fins (MFPF) was compared with metal foam heat sink and pin fin heat sink. Subsequently, the use of MFPF heat sink is simulated to remove non-uniform heat flux with different power levels. The non-uniform heat flux levels, Reynolds number, and the morphological parameters of metal foam are varied, and their effects on the heat transfer behavior examined.

2. Physical model description

The schematic diagram of the MFPF heat sink considered in this work, which is derived from the combination of metal foam heat sink and traditional pin fin heat sink, is shown in Fig. 1. . Along the flow direction, the heat sink is divided into three zones, an inlet metal foam region (L_1=28.75mm), the metal foam and pin fins combined region (L_2=22.5 mm), and an outlet metal foam region (L_3=28.75mm). The overall dimensions of the channel are L (80 mm)×W (35 mm)×H (10 mm). The dimensions of the computational domain are summarized in Table 1. Water is used as the coolant, and its thermo-physical properties are assumed constant. The fluid

enters the MFPF heat sink with a uniform velocity (u_{in}) and a constant temperature of 300 K. The top wall and the two side walls of the computational domain are adiabatic.

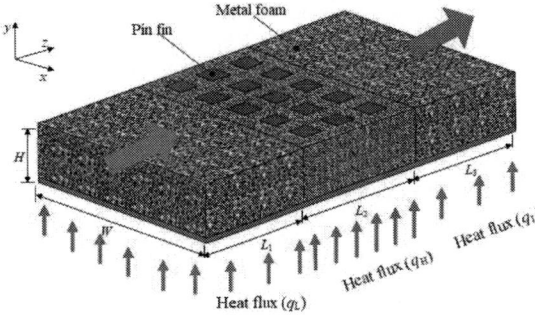

Figure 1: Schematic diagram of metal foam heat sink with pin fins

Table 1 The parameters used in the numerical simulation

Parameter	Value range	Unit
Porosity	0.8-0.95	1
Pore density ω	30-60	PPI
Length L_1,L_3	28.75	mm
Length L_2	22.5	mm
Channel width W	35	mm
Channel height H	10	mm
Thickness of substrate δ	0.5	mm
Low heat flux q_L	25	W/cm^2
High Heat flux q_H	25-100	W/cm^2

Bhattacharya et al. [18], DeGroot et al. [19], Seyf et al. [20], Feng et al. [21] and Jeng et al. [22] numerically and experimentally investigated the forced convective heat transfer in finned metal foam heat sink and found significant improvement in heat transfer compared to metal foam heat sink. In this work, the configuration of metal foam with pin fins is adopted to dissipate non-uniform heat flux for applications in electronics cooling. A copper plate with thickness of 0.5 mm is placed between metal foam and the heater with non-uniform heat flux at the bottom wall of the channel. A relatively lower heat flux is applied at the bottom wall of the inlet and outlet metal foam regions, and a higher heat flux at the bottom wall of the metal foam and pin fin combination region. In addition, the traditional pin fin heat sink and metal foam heat sink serve as baseline comparison models to evaluate the improvement in heat transfer and pressure drop by adopting the metal foam heat sink with pin fins.

3. Numerical details

In this section, the governing equations, the corresponding boundary conditions and numerical solution methods are introduced. The mesh independence analysis, as well as the model validation are also described.

3.1 Governing equation and boundary condition

The fluid flow in the computational domain is considered to be incompressible, laminar and steady. To simplify the numerical simulations, several assumptions are made:(1) The metal foam is homogenous and isotropic; (2) The thermo-physical properties of metal foam, fluid, and pin fins are constant; and (3) The metal foam and the fluid are in local thermal non-equilibrium (LTNE) state.

The conjugate heat transfer in the MFPF heat sink is numerically investigated by using the Forchheimer-Brinkman extended Darcy momentum equation, and the local thermal non-equilibrium energy equation. The heat transfer in pin fins is governed by the steady-state heat conduction equation. Accordingly, the conservation equations for mass, momentum and energy can be expressed as follows:

Continuity equation:
$$\nabla \cdot (\rho V) = 0 \tag{1}$$

Momentum equations:
$$\frac{\rho_f}{\varepsilon^2}(V \cdot \nabla)V = -\nabla P + \mu_{f,\text{eff}} \nabla^2 V - \left(\frac{\mu_f}{K} + \frac{\rho_f C_F}{\sqrt{K}}V\right)V \tag{2}$$

Energy equations:
Fluid phase:
$$(\rho c_p)_f (V \cdot \nabla T_f^f) = \nabla \cdot (k_{fe} \nabla T_f^f) + h_{sf} a_{sf}(T_s^s - T_f^f) \tag{3-1}$$

Solid phase:
$$\nabla \cdot (k_{se} \nabla T_s^s) - h_{sf} a_{sf}(T_s^s - T_f^f) = 0 \tag{3-2}$$

where ε is the porosity; T_f^f and T_s^s represent the temperatures of the fluid phase and solid phase, separately; k_{fe} and k_{se} denote the effective thermal conductivities of fluid phase and metal foam solid matrix, respectively, where $k_{fe}=\varepsilon k_f$ and $k_{se}=(1-\varepsilon) k_s$. K, C_F, h_{sf}, and a_{sf} are the permeability, inertial coefficient, interstitial heat transfer coefficient and specific area, respectively. The correlations of the pertinent parameters used in this paper is listed in Table 2.

For the simulations, the fluid enters the heat sink with a uniform velocity and a constant temperature. Non-uniform heat fluxes are applied at the bottom wall of the heat sink. The top and the two side walls of the computational domain are adiabatic. No-slip boundary conditions are applied at the channel walls and pin fin surfaces, and temperature and heat flux continuity are assumed at the interfaces of metal foam and pin fins.

3.2 Numerical method and mesh independent test

In this study, the fluid flow in the channel is assumed laminar and incompressible, and the governing equations (1)-(3) are solved based on the finite-volume method. To improve the numerical accuracy and computing efficiency, the entire domain is meshed with hexahedral meshes, which are refined near the pin-fin-to-metal-foam interfaces. The convective term in the momentum equations is discretized with the QUICK scheme, and the resulting algebraic equations are solved iteratively with the convergence criterion between two successive iterations less than 10^{-6}.

A mesh independence analysis is performed prior to performing the detailed numerical simulations. The computational domain of MFPF heat sink with $Re=500$ and $q_H=100$ W/cm^2, $q_L=0$, $\omega_H=30$PPI, $\omega_L=0$ is considered. Four sets of meshing systems containing total elements about (1) 754 thousand, (2) 933 thousand, (3) 1.7 million and (4) 2.3 million are tested. The maximum bottom temperature and the pressure drop are selected as the parameters for analysis. It is found that the variation of pressure drop is larger than maximum temperature, with refinement of mesh. The pressure drop deviations of Mesh 1 and Mesh 2 are larger than that of Mesh 3. For Mesh 3, the pressure drop and maximum temperature deviations of 0.39% and 0.006%, respectively,

are found. The grid system of Mesh 3 is considered to provide sufficient accuracy, without consuming excessive computing resources.

Table 2 Correlations of parameters for metal foam

Parameter	Correlation	Ref.
Pore size	$d_p = 0.0254/\omega$	[23]
Fiber diameter	$d_f = d_p \cdot 1.18 \dfrac{\sqrt{(1-\varepsilon)/(3\pi)}}{1-\exp((\varepsilon-1)/0.04)}$	[25]
Surface area	$a_{sf} = \dfrac{3\pi d_f \left[1-e^{-((1-\varepsilon)/0.04)}\right]}{(0.59 d_p)^2}$	[24]
Permeability	$K = 0.00073(1-\varepsilon)^{-0.224}(d_f/d_p)^{-1.11} d_p^2$	[25]
Inertial coefficient	$C_F = 0.00212(1-\varepsilon)^{-0.132}(d_f/d_p)^{-1.63}$	[25]
Heat transfer coefficient	$h_{sf} = \begin{cases} 0.76 Re_d^{0.4} Pr^{0.37} k_f/d, & (1 \le Re_d \le 40) \\ 0.52 Re_d^{0.5} Pr^{0.37} k_f/d, & (40 \le Re_d \le 10^3) \\ 0.26 Re_d^{0.6} Pr^{0.37} k_f/d, & (10^3 \le Re_d \le 2\times10^5) \end{cases}$ $d = \left(1-e^{-((1-\varepsilon)/0.04)}\right) \cdot d_f, \ Re_d = \rho_f ud/\mu_f$	[23]

Table 3 Variation of pressure drop and maximum bottom temperature at different grid sizes

Parameter	Pressure drop (Pa)	Maximum temperature (K)	Difference (%)	
Mesh 1	257.50	349.99	0.73	0.229
Mesh 2	256.84	350.01	0.99	0.017
Mesh 3	258.77	350.05	0.39	0.006
Mesh 4	259.40	350.07	Baseline	

3.3 Model validation

The present numerical method is validated by comparing the results with Feng et al. [21]. In their study, the forced convection heat transfer in metal foam and finned metal foam (FMF) heat sinks were numerically and experimentally investigated. The parameters used in their numerical simulation are as follows: metal foam porosity of 0.9, form

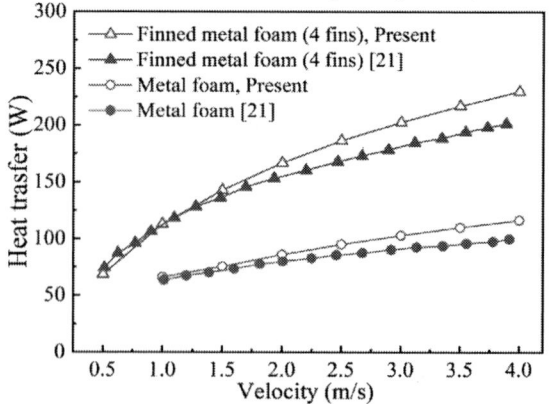

Figure 2: Comparison of the present numerical results with Ref. [21]

drag coefficient and permeability o f0.088 and 9.0×10^{-8} m^2, respectively. The metal foam heat sink and the finned metal foam heat sink with four fins of the same dimensions and boundary conditions are adopted for comparison. Fig.2 indicates that the present numerical results agree with Ref. [21], with a maximum discrepancy within 12%. The discrepancy may resulted from the thermal resistance between the foam and the fins, and inevitable properties difference between the real foam structure and the simplified structure in the simulation.

4. Results and discussion

In this section, the thermal effectiveness of MFPF heat sink is demonstrated by comparing the thermal performance of MFPF heat sink with metal foam heat sink and pin fin heat sink, wherein the computational domain of interest is only the middle section. Based on this, the implementation of MFPF heat sink for non-uniform heat flux electronics cooling is parametrically studied under different levels of non-uniform heat flux, Reynolds number, porosities, and pore densities.

4.1 Thermo-hydraulic performance comparison

Fig.3 shows the temperature distribution of the heated wall along the flow direction for different heat sink configurations, with heat fluxes of q_H of 25 W/cm^2 and q_L of 0 applied at the bottom wall. It is notable that, there is a significant increase of temperature along the flow direction for pin fin heat sink for both Re of 500 and 1000. However, the bottom wall temperature is dramatically reduced when metal foam is employed instead of pin fin, which implies that the insertion of metal foam has better heat transfer performance. In addition, as shown in Fig. 3 (c), the bottom wall temperature levels with metal foam and pin fin (MFPF) heat sink are considerably lower and more uniform, compared to the traditional pin fin heat sink and metal foam heat sink. It indicates that the pin fin and metal foam combined configuration is effective in heat removal.

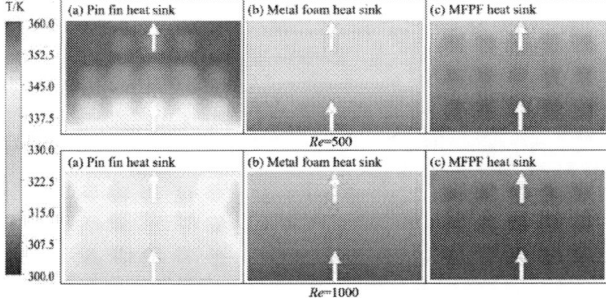

Figure 3: Comparison of bottom wall temperature distribution (a) Pin fin heat sink (b) Metal foam heat sink and (c) MFPF heat sink (ε_H=0.9, ε_L=1, ω_H=30 PPI, ω_L=0 PPI, q_H=25 W/cm^2, q_L=0)

The effects of heat sink configurations on the average Nusselt number at various Re are presented in Fig.4. In comparison with the traditional pin fin heat sink, the predicted results of metal foam heat sink and MFPF heat sink reveal a significant improvement in average Nusselt number. As expected, the average Nusselt number increases with Re, and the improvement becomes more pronounced when compared with the pin fin heat sink at relatively high Reynolds number. The implementation of metal foam enhances convective

surface area and intensifies fluid mixing capability, thereby improving heat transfer and average Nusselt number. Moreover, among the three heat sinks, the MFPF heat sink has the largest Nusselt number, 6.30 and 1.56 times compared to pin fin heat sink and metal foam heat sink at Re of 1000. The heat transfer enhancement of MFPF heat sink can be attributed to the addition of pin fins that promptly spread heat through the entire domain, which is then dissipated away from metal foam matrix by convective heat transfer. The above results confirm that the MFPF heat sink outperforms the pin fin heat sink and metal foam heat sink.

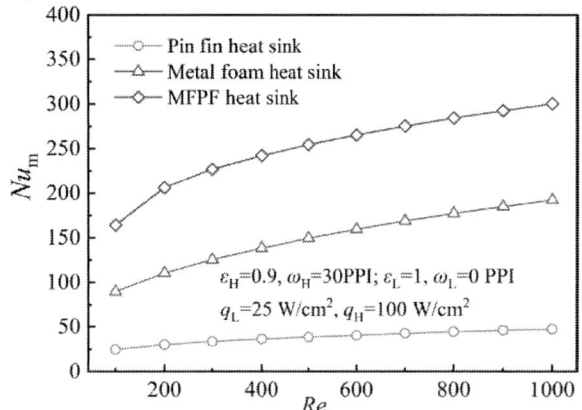

Figure 4: Comparison of average Nusselt number versus Re for various heat sinks

The design of MFPF heat sink greatly enhances heat transfer performance, however, due to the combination of pin fins and metal foam material, the pressure drop encountered is also important to consider. . Fig.5 shows the comparison of pressure drop with Reynolds number for the various heat sinks. The pressure drops exhibit a monotonic increase with Re, with the values for metal foam heat sink and MFPF heat sink being larger than that of pin fin heat sink. This is because metal foam has very complex structures with interconnected slender matrix, which greatly increases the flow resistance. It also indicates that the MFPF heat sink has the largest pressure drop, resulting from the common effects of impermeable pin fins and metal foam ligaments as obstacles block the flow passage. The pressure drop of MFPF heat sink is around 40.26 and 2.15 times of the pin fin and metal foam heat sink

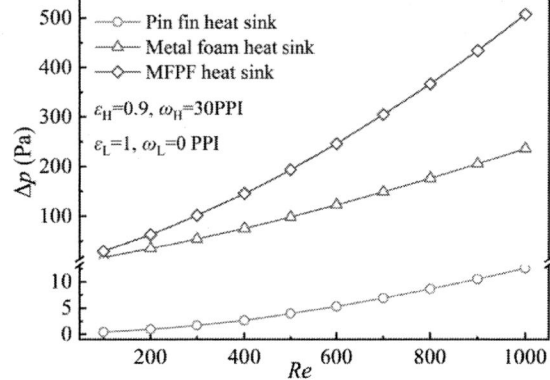

Figure 5: Comparison of pressure drop versus Re for various heat sinks

at *Re* of 1000. This implies that adding metal foam into the channel leads to a high pressure drop, as a result, more pumping power needs to be consumed to pump the fluid flow.

4.2 Parametric study of MFPF heat sink with non-uniform heat flux

MFPF heat sink demonstrates significantly enhanced heat transfer, which is more suitable for high heat flux dissipation. In this paper, the MFPF heat sink is employed for non-uniform heat flux electronic components cooling. Parametric study of MFPF heat sink is conducted by considering the effects of power levels of heat flux, fluid velocity, and metal foam morphological parameters (porosity and pore density (pores per inch, PPI).

4.2.1 MFPF heat sink with different power levels

The MFPF heat sink is employed to remove the dissipated heat from electronic components with non-uniform heat flux distribution, particularly applied with local high heat flux. A constant low heat flux (q_L) of 25 W/cm² is placed at the pure metal foam region, and the high heat flux region (q_H) is subjected at the metal foam and pin fin hybrid region with various power levels of 25, 50, 75, and 100W/cm². The bottom wall temperature distribution of MFPF heat sink subjected to different local high heat fluxes are presented in Fig.6. As the local heat flux increases, the temperature of hotspot region continues to rise. Local high heat flux of 100 W/cm² can be successfully removed, while maintaining the bottom wall temperature below 90°C by using the proposed MFPF heat sink.

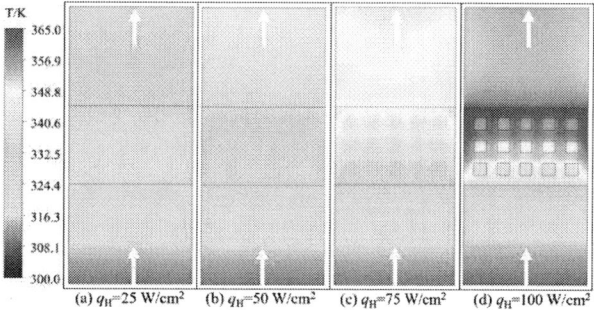

(a) q_H=25 W/cm² (b) q_H=50 W/cm² (c) q_H=75 W/cm² (d) q_H=100 W/cm²

Figure 6: MFPF heat sink with different local heat fluxes (*Re*=500, ε_L=ε_H=0.9, ω_L=ω_H= 30 PPI)

4.2.2 Effects of fluid velocity on thermal performance

MFPF heat sink as an efficient solution is promising for non-uniform heat flux dissipation, as demonstrated above, the non-uniform heat flux of 25,100 and 25 W/cm² placed at the bottom wall along the flow direction can be successfully dissipated with maximum junction temperature below 90 °C. The effects of fluid velocities (the corresponding *Re* is 200-1000) on the thermal performance of MFPF heat sink are investigated. Fig.7 presents the bottom wall temperature distribution with the heat fluxes of q_L=25 and q_H=100 W/cm² for different Reynolds numbers. It is noticeable that there is a significant temperature increase along the flow direction at lower *Re* condition, due to the high heat flux generated by the heat source and cannot be dissipated by small amount of fluid. However, the bottom wall temperature significantly deceases with the increase of fluid velocity resulting from the enhanced convective heat transfer. In particular, the bottom temperature

at the middle section is higher, because a high local heat flux is applied at this region to simulate the non-uniform heat flux.

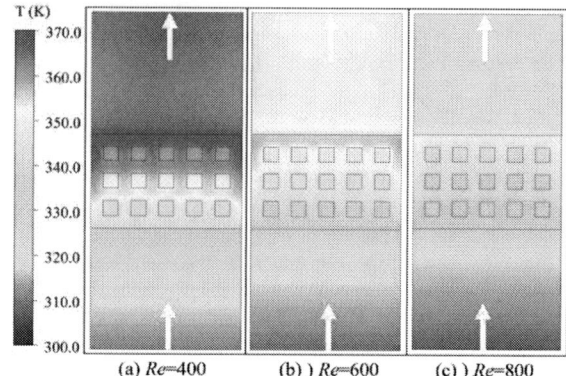

(a) *Re*=400   (b)) *Re*=600   (c)) *Re*=800

Figure 7: Bottom wall temperature distribution of MFPF heat sink versus *Re* (ε_L=ε_H=0.9, ω_L=ω_H= 30 PPI, q_L=25 W/cm², q_H=100 W/cm²)

Fig.8 plots the effects of Reynolds number on pressure drop and maximum temperature of MFPF heat sink. It is seen that the maximum bottom wall temperature deceases with the increasing *Re*. This is because convective heat transfer is enhanced as *Re* increases. The reduction of maximum temperature is approximately 58.8 K by increasing *Re* of 200 to 1000. However, the pressure drop of the MFPF heat sink exhibits an increasing trend with *Re*. Larger pressure drop means that more pumping power is required to pump the fluid. The pressure drop of MFPF increased by 611.9% as *Re* increases from of 200 to 1000. The results suggest that the MFPF heat sink can greatly enhance the heat transfer, however, a large pressure drop is required as a penalty, which may restrict its application in power limited engineering cases.

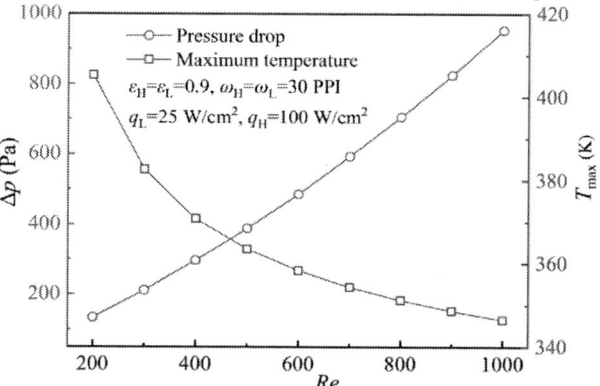

Figure 8: Effect of Reynolds number on pressure drop and maximum temperature of MFPF heat sink

4.2.3 Effects of pore density

In attempt to investigate the pertinent parameters on the thermal performance of MFPF heat sink, the non-uniform heat flux of 25,100 and 25 W/cm² is adopted. The thermal behavior of MFPF heat sink is also examined by changing the foam pore density in the high heat flux region, and the other region foam pore density remains 30 PPI as a constant value. As illustrated in Fig.9, increasing the pore density in the high heat flux region reduces the local temperature due to the increased convective heat transfer area. However, the reduction of temperature is slightly dependent on the variation of pore density, indicating that the thermal performance of MFPF heat sink is less sensitive to the variation of pore density.

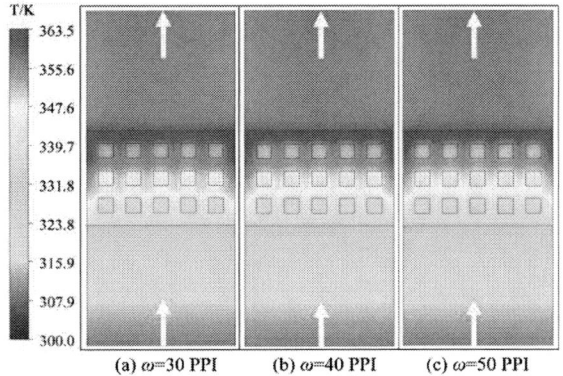

(a) ω=30 PPI   (b) ω=40 PPI   (c) ω=50 PPI

Figure 9: Effect of pore density on bottom wall temperature distribution of MFPF heat sink (Re=500)

The effects of pore density (ω_H) on the pressure drop and maximum temperature of MFPF heat sink are shown in Fig.10 at Re of 500. The pressure drop sharply rises as the pore density increases from 30 to 60 PPI. Decrease of pore density narrows the flow passage for coolant penetrating the porous medium, as a result, the fluid encounters a larger resistance as it flows though the internal porous medium. For MFPF heat sink, the configuration with foam pore density of 60 PPI in high heat flux region, the pressure drop increased by 101.9% compared to the MFPF heat sink with 30 PPI. The bottom wall maximum temperature in the high heat flux region shows the opposite trends, decreasing with the increase of pore density. Whereas, the variation of maximum temperature is not obvious with pore density, the reduction of maximum temperature of MFPF heat sink temperature is approximately 1 K. The results imply that the variation of pore density of MFPF heat sink has more pronounced effect on the pressure drop than heat transfer performance.

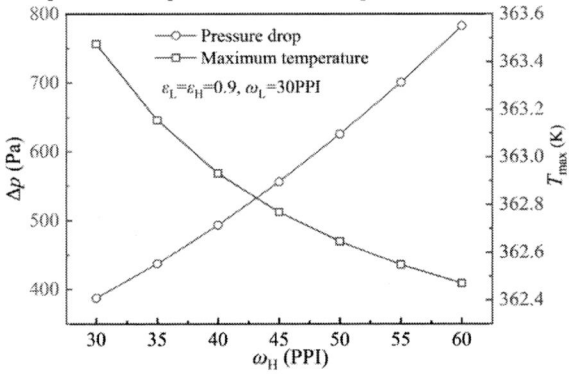

Figure 10: Effect of pore density on pressure drop and

4.2.4 Effects of porosity

The influence of metal foam porosity in the high heat flux region on the thermal performance of MFPF heat sink is examined with the non-uniform heat flux distribution of 25,100 and 25 W/cm². Fig.11 shows the bottom wall temperature distributions with the variations of porosities. It reveals that the bottom wall temperature, especially the high

heat flux region temperature, significantly increases when high porosity metal foam is employed. As low porosity metal foam is utilized, the temperature in the region is considerably reduced compared with high porosity. This suggests that the combination of pin fins and low porosity metal foam is more efficient to dissipate the high heat flux.

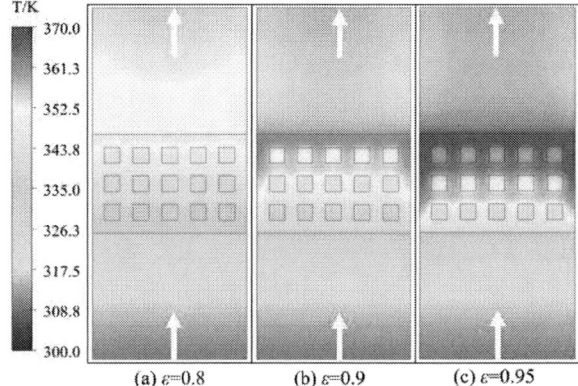

(a) ε=0.8   (b) ε=0.9   (c) ε=0.95

Figure 11: Effect of porosity on bottom wall temperature distribution of MFPF heat sink (Re=500)

The thermal performance of MFPF heat sink is also quantitatively evaluated by investigating the results of pressure drop and bottom wall maximum temperature with porosity, as shown in Fig.12. It indicates that increasing the porosity of metal foam in high heat flux region dramatically reduces the pressure drop. For example, the pressure drop of MFPF heat sink with porosity of 0.95 decreases by 25.0% compared with that of 0.8. The reduction of pressure drop results from the decreased flow resistance, i.e., the permeability increases with the increase of porosity. On the contrary, increasing porosity results in an increase in the maximum wall temperature. The maximum temperature of MFPF heat sink with porosity of 0.95 increases by 19.2 K compared with that for porosity of 0.8. This is attributed to the fact that increasing the porosity decreases both the effective thermal conductivity and convective surface area. As demonstrated in Fig.10, the pore density (increasing pore density increases the surface area) has little impact on the thermal performance of the MFPF heat sink. Therefore, the local temperature rise can be more effectively suppressed using MFPF heat sink with lower porosity.

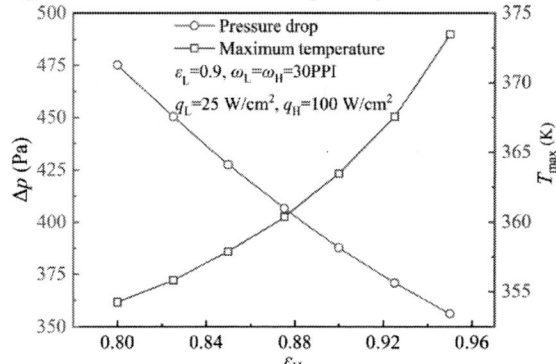

Figure 12: Effect of pore density on pressure drop and maximum temperature of MFPF heat sink

5. Conclusion

In this work, the performance of a metal foam and pin fin (MFPF) heat sink is simulated for single phase forced convection using water for thermal management of high powered electronics with non-uniform heat flux. The thermal effectiveness of MFPF heat sink is first demonstrated in comparison to the traditional pin fin heat sink and metal foam heat sink. The implementation of MFPF heat sink for non-uniform heat flux removal is investigated in terms of temperature distribution, pressure drop and maximum temperature. Based on above investigation, the following conclusions can be drawn:

(1) MFPF heat sink is more effective in suppressing the bottom wall temperature, compared to the traditional pin fin heat sink and metal foam heat sink, making it a promising solution for thermal management.

(2) MFPF heat sink provides a higher average Nusselt number due to the increased equivalent effective thermal conductivity with the combination of pin fin and metal foam. The average Nusselt number of MFPF heat sink performs up to 6.30 and 1.56 times compared to the pin fin heat sink and metal foam heat sink, respectively.

(3) The proposed MFPF heat sink can successfully provide thermal management of electronic device with non-uniform heat flux of 25,100 and 25 W/cm^2 placed along the flow direction at $Re=500$, while maintaining the bottom wall temperature below 90 ℃.

(4) The thermal performance of MFPF heat sink is strongly dependent on the metal foam morphological parameters (porosity and pore density). The heat transfer performance of MFPF heat sink is more sensitive to porosity than pore density, and the pressure drop is more sensitive to pore density than porosity.

Acknowledgements

The authors would like to acknowledge the financial support from the National Natural Science Foundation of China (No. 51676208). Chinese Scholarship Council (CSC) financially supported Yongtong Li to participate in the Joint Ph.D. Program (No. 201806450030) in the United States.

Reference

[1] Khoshvaght-Aliabadi M., Zangouei S., Hormozi F., "Performance of a plate-fin heat exchanger with vortex-generator channels: 3D-CFD simulation and experimental validation", International Journal of Thermal Sciences, Vol. 88, pp. 180-192, 2015.

[2] Zhao J., Huang S.B., Gong L., "Numerical studies on geometric features of microchannel heat sink with pin fin structure", Applied Thermal Engineering, Vol. 93 pp.1347-1359, 2016.

[3] Zhai Y.L., Xia G.D., Liu X.F., et al, "Heat transfer in the microchannels with fan-shaped reentrant cavities and different ribs based on field synergy principle and entropy generation analysis", International journal of heat and mass transfer, Vol.68, pp.224-233, 2014.

[4] Han X.H., Wang Q., Park Y.G., et al, "A review of metal foam and metal matrix composites for heat exchangers and heat sinks", Heat Transfer Engineering, Vol.33, No.12, pp. 991-1009, 2012,.

[5] Rachedi R., Chikh S., "Enhancement of electronic cooling

by insertion of foam materials", Heat and Mass Transfer, Vol.37, No.4-5, pp. 371-378, 2001.

[6] Bayomy A.M., Saghir M.Z., Yousefi T., "Electronic cooling using water flow in aluminum metal foam heat sink: Experimental and numerical approach", International Journal of Thermal Sciences, Vol.109, pp.182-200, 2016.

[7] Singh R., Akbarzadeh A., Mochizuki M., "Sintered porous heat sink for cooling of high-powered microprocessors for server applications", International Journal of Heat and Mass Transfer, Vol.52, No.9-10, pp. 2289-2299, 2009.

[8] Wan Z.M., Guo G.Q., Su K.L., et al, "Experimental analysis of flow and heat transfer in a miniature porous heat sink for high flux application", International Journal of Heat and Mass Transfer, Vol. 55, No. 15-16, pp.4437-4441, 2012.

[9] Hung T.C., Huang Y.X., Yan W.M., "Thermal performance of porous microchannel heat sink: effects of enlarging channel outlet", International Communications of Heat and Mass Transfer, Vol. 48 pp. 86–92, 2013.

[10] Hung T.C., Huang Y.X., Yan W.M., "Thermal performance analysis of porous-microchannel heat sinks with different configuration designs", International Journal of Heat and Mass Transfer, Vol. 66, pp. 235-43 2013.

[11] Boomsma K., Poulikakos D., ZwickF., "Metal foams as compact high performance heat exchangers", Mechanics of materials, Vol.35, No.12, pp.1161-1176, 2003.

[12] Li Y.T., Gong L., Xu M.H., et al, "Thermal performance analysis of biporous metal foam heat sink", Journal of Heat Transfer, Vol.139, No.5, 2017,.

[13] Jiang P.X., Li M., Lu T.J., et al, "Experimental research on convection heat transfer in sintered porous plate channels", International Journal of Heat and Mass Transfer, Vol. 47, No.10, pp. 2085-2096, 2004.

[14] Hadim A., "Forced convection in a porous channel with localized heat sources", Journal of heat transfer, Vol.116, No.2, pp. 465-472,1994.

[15] Yang J., Zeng M., Wang Q.W., et al, "Forced convection heat transfer enhancement by porous pin fins in rectangular channels", Journal of Heat Transfer, Vol. 132, No.5, pp.051702, 2010,.

[16] Sener M., Yataganbaba A., Kurtbas I., "Forchheimer forced convection in a rectangular channel partially filled with aluminum foam", Experimental Thermal and Fluid Science, Vol. 75, pp.162-172, 2016.

[17] Farsad E., Abbasi S.P., Zabihi M.S., "Fluid flow and heat transfer in a novel microchannel heat sink partially filled with metal foam medium", Journal of Thermal Science and Engineering Applications,Vol.6, No.2, pp. 021011, 2014.

[18] Bhattacharya A., Mahajan R.L., "Finned metal foam heat sinks for electronics cooling in forced convection", Journal of Electronic Packaging, Vol.124, No.3, pp.155-163, 2002.

[19] DeGrootC .T., Straatman A.G., Betchen L.J., "Modeling forced convection in finned metal foam heat sinks",

Journal of Electronic Packaging, Vol.131, No.2, pp.021001, 2009.

[20] Seyf H.R., Layeghi M., "Numerical analysis of convective heat transfer from an elliptic pin fin heat sink with and without metal foam insert", Journal of Heat Transfer, Vol.132, No.7, pp. 071401, 2010,.

[21] Feng S.S., Kuang J.J., Wen T., et al, "An experimental and numerical study of finned metal foam heat sinks under impinging air jet cooling", International Journal of Heat and Mass Transfer, Vol. 77, pp. 1063-1074, 2014,.

[22] Jeng T.M., Tzeng S.C., Tang F.Z., "Fluid flow and heat transfer characteristics of the porous metallic heat sink with a conductive cylinder partially filled in a rectangular channel", International Journal of Heat and Mass Transfer, Vol.53, No.19, pp. 4216-4227, 2010.

[23] Lu W., Zhao C.Y., Tassou S.A., "Thermal analysis on metal-foam filled heat exchangers. Part I: Metal-foam filled pipes", International journal of heat and mass transfer, Vol.49, No.15-16, pp.2751-2761, 2006.

[24] Calmidi V.V., Mahajan R.L., "Forced convection in high porosity metal foams", Journal of Heat Transfer, Vol.122, No.3, pp.557–565, 2000.

[25] Calmidi V.V., "Transport Phenomena in High Porosity Metal Foams", Ph.D. thesis, University of Colorado, Boulder, 1998.

Mechanical Cycling Reliability Testing of Thermal Interface Materials for Semiconductor Test

David L. Saums*[1], Tim Jensen[2], Carol Gowans[2], Ron Hunadi[2], Mohamad Abo Ras[3]

[1]DS&A LLC, Collaborative Innovation Works, 11 Chestnut Street, Amesbury MA 01913 USA
[2]Indium Corporation, 34 Robinson Road, Clinton NY 13323 USA
[3]Berliner Nanotest und Design GmbH, Volmerstrasse 9B, D-12489 Berlin, Germany

* Corresponding Author and Presenter; E: dsaums@dsa-thermal.com

Abstract

The most challenging applications for thermal interface materials where durability and thermal performance are both required is found in the semiconductor test and burn-in market. There are a wide range of application requirements given the different types of test sockets, test heads, and test equipment configurations. However, a dominant characteristic is the need for a single TIM, applied to a test device, to contact and release cleanly and without damage to either the TIM or to the device under test (DUT), ideally with the ability to conduct such cycles hundreds and thousands of times before replacement of the TIM is required. This places a significant burden on material design, to achieve substantial durability in harsh usage circumstances, while also providing a high level of thermal performance. A mechanical reliability test program for evaluating durability as well as thermal resistance performance of specialized thermal interface materials (TIMs) has been developed. These specialized TIMs have been developed specifically to meet requirements for semiconductor test and burn-in requirements, which are extreme; cycling with multiple contacts for a single TIM (up to thousands of cycles) is a long-sought development goal for the semiconductor test equipment industry.

An automated, servo-driven TIM test stand developed to follow industry-standard test methodology is used to conduct testing under controlled test conditions. Examples of test requirements are given, as well as a description of an automated contact cycling test designed to test per those requirements, with results for a set of newly-developed TIMs explicitly adapted for semiconductor test usage. Four test phases have been developed with increasingly challenging test requirements for durability while also exhibiting excellent thermal performance.

Keywords

Thermal resistance, thermal interface, semiconductor test, durability, cycling, device under test

Nomenclature

DUT   Device under test
OSAT   Out-Sourced Assembly and Test
Rth   Thermal resistance
TIM   Thermal interface material

1. Introduction

Thousands of thermal interface materials are manufactured, designed for an exceptionally wide range of application requirements in semiconductor and electronic systems. Development of a new TIM by a manufacturer requires identification of application requirements that are specific to a given application; this has become much more prevalent in recent years as the different types of applications have continued to expand. A critical requirement prior to beginning material development is the identification of specific attributes desired, as well as identification of the material characteristics that may be required to meet those attributes. The primary function of any TIM is to transfer heat from one surface, typically either metal or a semiconductor material such as silicon, to the surface of another component that will act to dissipate a given heat load. The early development of the TIM market saw wide use of relatively simplistic, generalized materials. Requirements today are substantially more specific and have tightened significantly, and that has led to identification of material categories (so-called thermal compounds, gap-filling materials, phase-change, graphitic, adhesive, metallic, polymer solder hybrids, and other categories are examples) and thermal performance values that may vary within and between categories.

Examining targets for the highest-performing TIMs typically focuses on development of exceptionally thin compounds and sheet forms that will compress and thin under load, for the simple reason that contacting surfaces are typically materials with much higher bulk thermal conductivity value. An overriding characteristic of the majority of TIM applications is installation of the TIM once, followed by completion of the assembly of the heat sink or other device to the semiconductor and TIM as a one-time process. A requirement for factory or field rework is the only typical explanation for the joined TIM to potentially be disassembled, disrupting the clamped interface condition. [Where a rework process is specified, standard protocol is that the mating surfaces would be cleaned and a new TIM applied, prior to reassembly, to ensure seating and performance of the interface material.]

2. Requirements for TIM for Semiconductor Test

Semiconductor test and burn-in applications represent a different and more complex situation. The TIM required is one which must survive multiple cycles of contact with a mating surface and release, without flaking or other detritus or marking of any kind, and this is the reason that semiconductor test may be considered to constitute, the most challenging requirements for TIM performance *and* durability. The semiconductor test industry has sought for years to identify a material that will survive multiple insertions of a test head with single TIM attached, contacting and releasing from hundreds or thousands of semiconductor devices under test. [1.] This is a very different scenario than is typical, for example, with the use of a dispensed and screened thermal compound or phase-change TIM in a specified thickness to a semiconductor, with heat sink placement once as a permanent assembly [again, excepting rework requirement(s)].

Additional complexity for TIM development and use in semiconductor test is found in the goals for high unit through-put testing, especially for semiconductor manufacturers and packaging companies (including out-sourced assembly and test companies, or OSAT). The primary example here is the ability to utilize a single test system with test head and placed TIM to contact different devices under test – bare die packages, lidded packages, packages with different contact surface (lid or die) sizes and configurations. This requirement has led to design of test systems that utilize a gimbal-mounted test head, able to swivel and rotate within a defined range to contact different package types and contact surfaces. This design concept leads to a potential non-flat condition for test head contact and what is termed as the "strike angle", wherein the TIM applied to the test head contacts the edge of a silicon die or edge of a package lid. In addition, test procedures may also include operation at temperatures to 115-125°C for defined periods, as burn-in; in certain tests, up to 155°C. Combined with non-parallel contacting surfaces and a potential strike angle as described, these mechanical test requirements place significant challenges on durability for a selected TIM.

Further, a test head and attached TIM may contact devices with different surface flatness and roughness conditions. This is again due to testing protocols whereby different semiconductor package configurations may be tested in a single test system, the purpose being to maximize test through-put and yield while minimizing system complexity and downtime between different batches of devices to be tested. The TIM applied to a test head may therefore contact a the backside of silicon die for a batch consisting of bare-die packages, followed by subsequent lots of a lidded package, such as an ASIC, and different lots may have different lid thicknesses, package heights, an lid or die footprint dimensions. All of these examples may also represent different surface roughness and flatness conditions.

The requirement for adequate minimum thermal resistance and minimized material thickness must apply in all cases as the baseline thermal performance criteria for the TIM.

Key factors in semiconductor test are various electrical performance tests for device performance binning identification. Elevated temperatures during burn-in testing are intended to stress and cull early failures.

[It is important to note that some electrical test systems utilize pin probe arrays to contact semiconductor devices. These pin probe array systems are not designed to utilize a TIM and therefore are an exception to this discussion.]

The common test head design that utilizes a TIM contains with heating and cooling capabilities within the test head. The flat surface of the test head contacts the DUT and rapidly applies cooling and heating cycles per a proscribed test regimen; the test head may incorporate active cooling (a liquid cold plate, thermoelectric modules, or both) as well as heaters, used to apply the specified test temperatures.

We refer to these requirements in total as requirements for semiconductor test. The test program described below has been designed primarily to determine if newly-developed TIMs, intended to meet these semiconductor test requirements, can survive an unusual mechanical durability regimen. This test program utilizes an automated TIM test stand because of the features designed into the standard test platform; the primary output in this project is not intended to be thermal resistance analysis.

More typical evaluation of a TIM involves traditional primary selection criteria: minimized thermal resistance under specified application conditions, suitability of the dispensing or placement method required, and ability of the TIM to operate without degradation for the required product assembly life. Development of a new TIM type therefore focuses first on tested TIM thermal performance; second, the requirement for the type of dispensing or assembly placement required, and third, the ability of the TIM to operate reliably once assembled over the expected product life without degradation. There is no requirement for repeated contact cycling. Only in the case of a component failure is there a recommended rework process that will include removal of the TIM, surface cleaning, and placement of a new TIM.

In summary, few TIM types are capable of meeting such challenging and unusual requirements, and no traditional concept of a TIM is expected to be contacting and releasing repeatedly from heat source surfaces. Very specialized materials have therefore been developed by a small number of companies, to attempt to meet these conditions. A test program is required to analyze the durability of newly-developed specialized TIMs.

3. Industry Requirements Survey for Semiconductor Test

A survey of manufacturers of semiconductor test industry socket and test equipment manufacturers and semiconductor manufacturers was conducted, to establish a set of test parameters. Assistance from test industry vendors was very important in order to develop a test regimen that would closely mimic the actual test system operating conditions. Each company performing semiconductor test has different procedures; there is also no uniform number of test cycles for which a single TIM is utilized, from company to company. The goal for the number of test cycles for a single TIM applied to a test head ranges from a single contact (for extremely high value devices) to many thousands; one test equipment manufacturer has searched for years for a single TIM able to survive tens of thousands of cycles for very high volume, high through-put operations. The purpose of the industry survey, therefore, was to develop a set of targets for

Table 1. Thermal/Mechanical Cycling Test Parameters			
Organization	Test Pressure Reported	Test Temperature Range Reported (°C)	Dwell (Seconds)
Company A	11.7 bar (170 PSI)	25**/100	60
	11.7 bar (170PSI)	100	60
Company B	6.7 bar (100 PSI)	-	60
Company C	-	120	-
Company D	-	100	-
Company E	-	80	60
Company F	4.1/6.7 bar (60/100 PSI)*	105**/125	-
	6.7 bar (100 PSI)*	105**/125	-

Notes: * Pressure applied dependent upon die or package contact area. ** Initial value.

Table 1. Mechanical cycling test parameters

typical device contact surface dimensions, force applied, test head (DUT) temperature, dwell time, and number of cycles to be performed. The strike angle and elevated temperature values (for semiconductor burn-in) were additional requirements identified in this survey. Values for survey questions are shown in Table 1.

4. Test Program Design

Based on survey results, a program was designed for testing a set of newly-developed TIMs. A commercial test stand with servo motor-driven test heads was selected, as the well-designed control program would allow automation of the cycling routine with tight controls and the ability to generate a set of output values, useful for control purposes. Table 2 summarizes the four test program phases. The initial phase was designed to establish baseline values, to indicate if one or more of the three materials to be tested could, under the least-rigorous conditions in this test program, achieve one thousand contact cycles without material failure. Material failure was to be determined by visual examination of the TIM, upon completion of one thousand contact-and-release cycles. The system measured thickness change for each cycle, to be used to determine that the test routine was stable; the thermal resistance value was seen as potentially an indicator of material failure, which would be determined by increasing thermal resistance values. Note that this is not intended to be a standard test for thermal resistance values, as for the introduction of the strike angle in Phases II and III would be an unusual and inappropriate method for simple thermal resistance testing; this would not represent the normal application condition with parallel contact surfaces for which a TIM is designed to be used. In addition, the sixty-second dwell time would normally be considered to be insufficient for system heat load stabilization and accurate data generation. Therefore, thermal resistance values were potentially a secondary indicator of material failure in this specialized testing.

The use of a commercially-available thermal interface material test stand designed to follow ASTM D 5470-17 that utilized servo motor-driven upper and lower test heads was

Table 2. Thermal/Mechanical Cycling Test Program Design				
Program Phase	Purpose	Test Head Configuration*	Operating Temperature (°C)	Data Output
I	Baseline Values	Parallel	70 – 95	R_{th}**, Thickness Change,*** 1,000 Contact Cycles
II	Strike Angle	Upper Body: Strike Angle	70 – 95	R_{th}**, Cycle Count
III	Strike Angle/Elevated Temperature	Upper Body: Strike Angle at Elevated Temperature	125	R_{th}**, Cycle Count
IV	Baseline Values	Parallel	95	R_{th}**, Thickness Change,*** 5,000 Contact Cycles

Notes: * Test head configuration and test system design per ASTM D 5470-17 test methodology. [2]
           ** Thermal resistance and (***) thickness values used to indicate a stabilized test routine.

Table 2. Thermal/mechanical cycling test program design.

Table 3. Thermal/Mechanical Cycling Test Head Design	
Property	**Value**
Material	Aluminum Alloy (AlMgSi1)
Contact Area	17.5mm x 17.5mm (306mm^2)
Contact Surface Roughness	Rz ≤ 1µm
Sample Temperature	95°C
Upper Reference Body (Heater Bar)	125°C
Lower Reference Body (Liquid Cold Plate)	75°C
Temperature Measurement	In situ
Thickness Measurement Under Force Applied	In situ

Table 3. Test head and operating temperatures utilized.

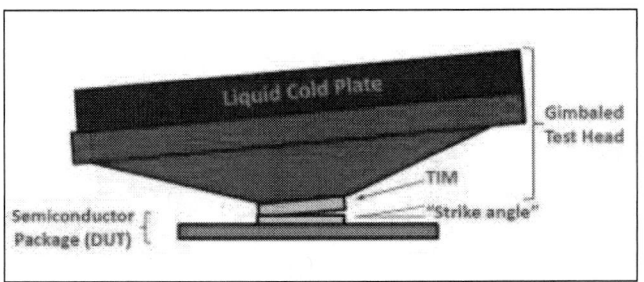

Figure 1. Illustration of a semiconductor test stand gimbal-mounted test head, allowing contact of the test head with attached TIM at different orientations to different devices under test with different package configurations and contact surface dimensions. (Adapted from Sanchez, [1]).

decided upon as the solution that would allow automation of the repetitive contact cycling, for a large number of exactly equivalent motions. The principals of this test methodology are available by purchase from ASTM International and discussion may be found in the literature. [2]

The selected test stand met these requirements and also allowed independent selection of upper and lower test head temperatures, incorporating a heating device for one and a chiller and liquid cold plate for the second. Incorporated within the design of the test stand are controls to allow selection of individual temperatures (upper and lower test heads), contact pressure to be applied, dwell (for the period of intimate contact), and the ability to select from a wide range of test head configurations and materials. This system provided output values that include temperature measurement, thickness measurement and change in thickness (for the selected TIM under test), bulk and interfacial thermal resistance values, and derived thermal resistance (Rth). [3, 4.] Additional features that included removable test heads allowed specification of a contact surface roughness for each test head; a fine adjustment feature allowed selection of parallel contact surfaces or the specification of an angled contact and a horizontal contact adjustment for one test head. This ability to specify the positioning and angle of one test head was critical for Phases II and III. [5.]

Specifications for the test heads are shown in Table 3. The contact area selected is intended to mimic the footprint dimensions of a lid used for a typical IC package. The temperatures for upper and lower test heads were selected to generate a sample temperature as a representation of the values indicated in the industry survey of testing parameters. Contact test pressure and dwell time were also selected based on the industry survey results. These features in the test stand design allowed this test program to proceed without the need for specially-designed equipment, an added cost and time penalty.

5. Materials Tested

A set of three newly-developed TIMs designed to meet the semiconductor test industry requirements identified above were submitted to an independent test laboratory for this test program. These three materials are described briefly in Table 4 (below), with a key for the test graphs that follow. Each is a metal alloy, in two cases with a second metal alloy applied during the manufacturing as a cladding; the third is a dead-soft aluminum alloy with a thermally-conductive coating applied to one surface only. The use of an aluminum alloy cladding on one surface of an indium alloy is designed to meet a very

Table 4. Thermal Interface Materials Tested	
Graph Key	**Description**
CLAD	Indium (99.99%) flat foil, clad one side (0.1µ/0.0005") aluminum
CLAD HSK	Indium (99.99% foil, clad one side (0.1µ/0.0005") aluminum, HSK pattern applied*
HSMF-OS	Aluminum foil (0.002"), coated one side with dry thermal compound**

Table 4. Newly-developed TIMs for semiconductor test requirements with basic descriptions and graph key, as tested herein. Additional information may be found in the reference.

specific requirement in practice: the ability of the TIM (under temperature and pressure conditions, attached to the test head) to separate cleanly from the DUT without leaving residue on the DUT surface. Indium metal is naturally relatively compliant and can be moderately tacky; the very thin aluminum alloy cladding provides a more durable and stiffer contact surface that ensures that no residue remains and the TIM will separate cleanly even after elevated temperature and pressures are applied. Interestingly, this manufacturing concept applied to a metallic TIM has also been significantly useful for high value semiconductor modules that may be subject to rework and field replacement with upgrades; examples are found in enterprise server processors. [6.]

The third material developed for use in semiconductor test

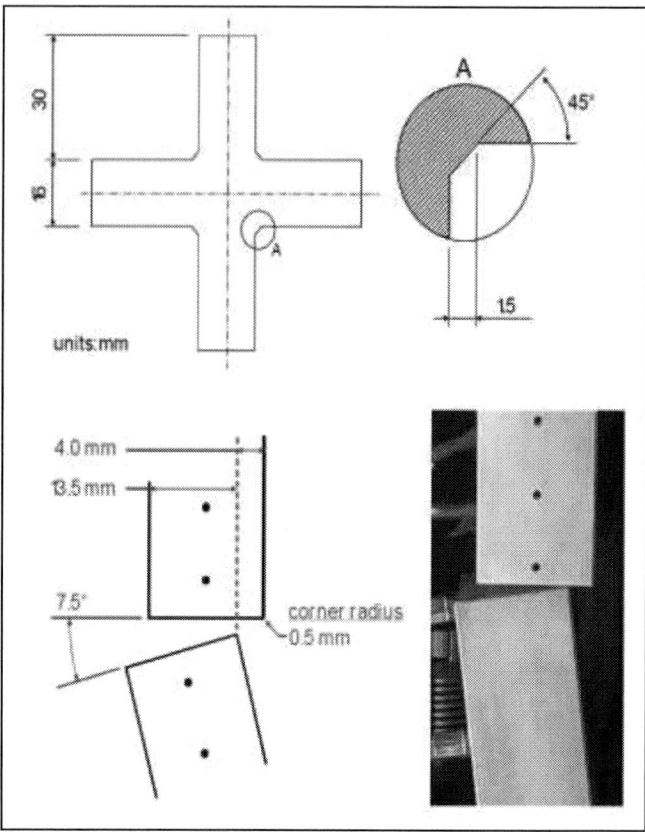

Figure 2. Test material applied to test head, Phase I. This is the HSMF-OS TIM die-cut to a so-called "Red Cross" shape and prior to completed assembly. The four arms will be folded against the sides of the upper test head and a simple mechanical fixture attached to clamp the TIM to the upper test head, to mimic placement in a semiconductor test system.

Figure 3. An example of the dimensioning of a so-called "Red Cross" die-cut TIM designed for use in semiconductor test systems The lower figure and photograph illustrate the test program design incorporating a 7.5° strike angle; the lower head moves in this implementation, to act as the contacting head.

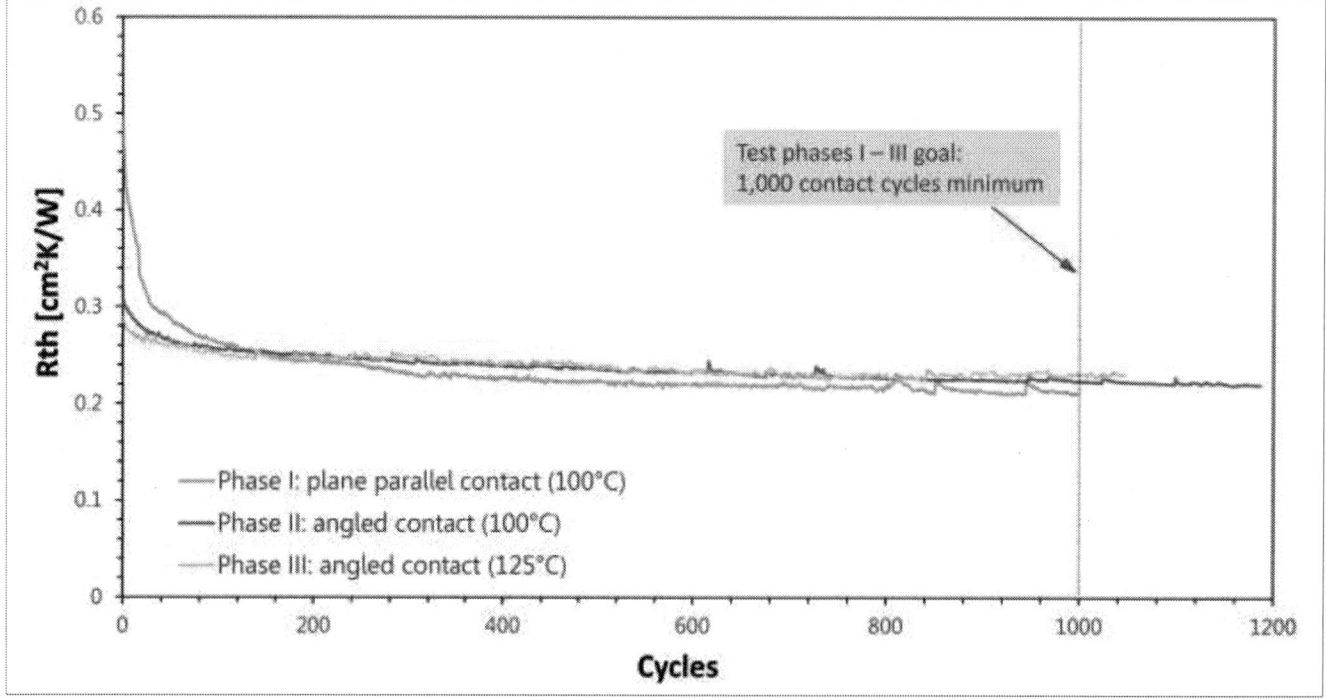

Figure 4. Thermal resistance values, Phases I - III, showing test results for 1,000 cycles completed per phase for a single material type. A change in attachment fixture resulted in lower initial values for Phases II and III, compared to Phase I.

42

is a dead-soft aluminum alloy foil that has a non-silicone thermal compound applied to one surface only. The purpose is again to address the requirement that the TIM consist of no polymeric material which will leave either a mark or any residue on the surface of the DUT. Recognize that the DUT may be either a smooth metal lid surface with a specified finish and manufacturer markings, or the backside of a semiconductor die; the requirement for no marking of the DUT and no residue is purely cosmetic but critical.

An example is shown in Figure 2 of one TIM during the process of attachment of the die-cut "Red Cross" TIM to the upper test head. RTDs are also evident, prior to insertion into the upper and lower test heads (requiring addition of through-holes at specified locations in the TIM).

6. Test Results and Analyses

Tested resistance values for Phases I - III of the test program are shown in Figure 4, for one of three selected materials... All three materials surpassed the test program goal of 1,000 contact cycles per phase.

A graph of thickness change for each material was also plotted from test data generated during cycling and showed a stable test condition.

Observations of test head surfaces showed no visible marking and zero residues, upon completion of each phase of each test for each material.

Successful conclusion of these three phases led to the addition of one additional phase, to determine if such metallic TIMs could successfully be tested in identical conditions to achieve 5,000 contact cycles. The HSMF-OS type was selected and tested and achieved this stretch goal, again with no residue remaining on the test head that mimicked the DUT and no marking of that surface.

7. Conclusions

A survey of requirements for an unusual set of application parameters for thermal interface materials was conducted, a test program was designed based on survey results, an automate test platform was utilized to generate mechanical contact cycling as a durability test for newly-developed metallic TIMs. Test data under four different sets of conditions, intended to mimic requirements found in the semiconductor test and burn-in industry, are shown that the selected TIMs successfully met each durability goal for these four test phases. Results have been analyzed and the industry requirements for no visible tearing, separation, or marking of the DUT have been met successfully.

Acknowledgments

All testing was conducted in an independent test laboratory, at Berliner Nanotest und Design GmbH, in Berlin, utilizing a test stand developed by Berliner Nanotest. Industry inputs by Jaime Sanchez, Intel Corporation, and other participants in the industry requirements survey, were very important for practical test program design.

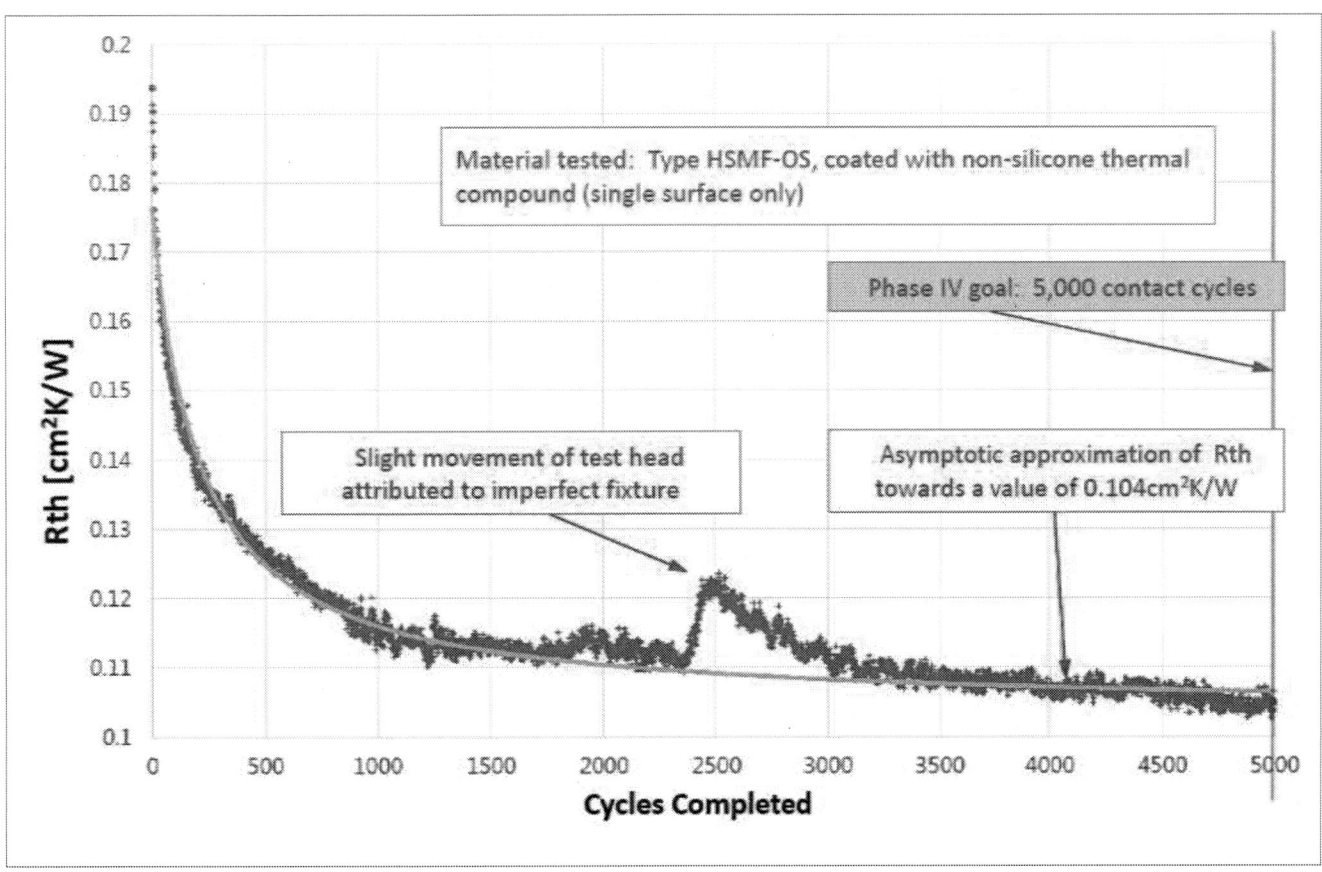

Figure 5. Thermal resistance values, Phase IV, showing test results for the stretch goal of 5,000 cycles completed for a single material type. A slight perturbation was analyzed; the reliability cycling test was completed successfully for this material.

References

1. Sanchez, J.A., Intel Corporation, "Challenges of Thermal Interface Materials in Test of IC Packages," IMAPS Advanced Technology Workshop on Thermal Management 2013, Los Gatos CA USA, November 5-7, 2013.

2. ASTM International, ASTM D 5470-17, Standard test method for thermal transmission properties of thin thermally conductive solid electrical insulation materials, ASTM International (Philadelphia PA USA, 2017).

3. Lasance, C., Murray, C.T., Saums, D.L., Rencz, M., "Challenges in Thermal Interface Material Testing," Proceedings, Semi-Therm Symposium 23, Dallas TX USA, March 2006.

4. Jarett, R.N., et al., "Comparison of Test Methods for High Performance Thermal Interface Materials," Proceedings, Semi-Therm Symposium 23, San Jose CA USA, March 2007.

5. Berliner Nanotest und Design GmbH, "TIMA 5 Thermal Interface Material Analyzer Data Sheet," August 2018, www.nanotest.eu/tima5.

6. Saums, D., Jensen, T., "Development, Testing, and Application of Metallic TIMs for Harsh Environments and Non-Flat Surfaces," Proceedings, IEEE I-Therm Conference 2017, Orlando FL USA, May 30-June 2, 2017.

High Performance Lightweight Ceramic Material for Thermal Management in Electronic Devices

Bei Xiang, Chandra Raman and Xiang Liu
Momentive Performance Materials Quartz, Inc.
22557 W. Lunn Road, Strongsville, OH 44149
Email: bei.xiang@momentive.com

SUMMARY

High performance thermally conductive, light-weight and low coefficient of thermal expansion materials are desired in novel thermal management designs for dissipating heat more efficiently in today's high power density electronic devices. Hexagonal boron nitride (h-BN) is a unique synthetic light-weight high performance thermally conductive ceramic material with excellent electrical insulation. h-BN has proven to be one of the most attractive fillers for improving thermal conductivity while maintaining the electrical insulation of polymeric thermal interface materials (TIM) and thermally conductive plastics (TCPs) used in electronic devices.

Due to the better isotropic thermal performance, agglomerate BN particle is preferred over platelet BN particle as thermal filler in polymeric TIM for high through-plane thermal conductivity. The impact of compounding conditions on the morphology of agglomerate BN, which is critical to final composite thermal performance, is discussed in this presentation. It is demonstrated that the final thermal conductivity of polymer composites filled with the same percentage of agglomerate BN could have 4x difference in thermal conductivity depending on compounding conditions.

While plastics are traditionally relatively poor conductors of heat, it is possible to significantly increase the thermal conductivity with the suitable use of additives. BN offers the great solution to enhance thermal conductivity while maintaining electrical insulation and keeping full freedom in the color space. This presentation will explore the possible uses of thermally conductive plastics (TCPs) to solve thermal management challenges in a broad range of applications. TCPs could potentially replace aluminum as the material of choice for heat sinks. This discussion will involve both theoretical and finite element modeling approaches to compare the thermal performance of aluminum and TCP heat sinks, under both free- and forced-convection environments. The results show that thermally conductive plastics could potentially replace aluminum in free-convection environments, while delivering additional benefits over standard die-cast aluminum heat sinks with significantly reduced weight at little to no additional cost on a per part basis.

1. INTRODUCTION

As more and higher power chips are incorporated into smaller electronic devices nowadays, power density has increased tremendously. The heat generated within the shrunk space in these electronic devices must be dissipated more efficiently, or else the heat buildup could quickly throttle the device performance and even result in failure. The miniaturization of many mobile electronic devices makes active cooling more often not an applicable choice for thermal management design in current and future devices. The desire for enhanced passive heat dissipation solution entangled with the requirement for weight reduction of the whole package forces thermal designers to look for novel high performance lightweight thermally conductive materials, which are used in different components of the electronic devices, such as thermal interface material (TIM), housing materials, heat sink fins and heat spreaders [1].

Thermoset polymers, such as silicone and epoxy, have been extensively used for preparing thermal interface materials for decades. Because of the thermally insulative nature of polymers (TC ~0.2 W/mK), ceramic fillers like Al_2O_3, ZnO, etc. have been used for improving the thermal conductivity of polymer composites. However, the thermal performance criteria of TIM are increasingly demanding, and TIM makers are facing great challenges to achieve the thermal conductivity satisfying the thermal management requirement of today's high performance electronic devices. h-BN is a superior thermal filler with the highest in-plane thermal conductivity (>300 W/mK) and lowest dielectric constant in the typical ceramic fillers found in polymer composites. However, the anisotropy in h-BN's thermal conductivity has greatly limited the thermal performance of the TIM made of BN-polymer composite in the direction needed. To solve this issue, agglomerate BN, whose particles are the random agglomeration of fine platelets, has been developed. Because of the random orientation of the platelets in the agglomerates, the thermal performance in all directions are very close if not equal. Significant improvement of all-direction TC has been achieved for polymer TIM filled with agglomerate BN [2]. However, the optimal thermal and rheological performance is often not obtained by (new) users because of the lack of understanding on the impact of compounding conditions on BN morphology, which directly influence the thermal and rheology performance of BN filled polymer composites. Therefore, a study is performed on correlation among compounding conditions, BN morphology, and final polymer composites' thermal performance.

Thermoplastics have been used extensively in automobile, LEDs, computers, and other consumer electronic devices to enable lighter and cheaper products. Thermal management parts like heat sinks remain to be the area where thermoplastics have not been able to replace metals due to its inherent electrical insulation property. BN has been used successfully as thermal filler in making thermally conductive electrically insulative plastics composites. Although BN filled TCPs, which typically have 1-10 W/mK thermal conductivity, is yet as thermally conductive as metals such as aluminum with typical thermal conductivity of 80-150 W/mK, the heat dissipation effect of parts made of TCPs could still compete against the parts made of metals in the environments where convection is limited. In the case study reported in this presentation, the cooling effects of heat sinks made of TCPS are compared with the aluminum counterparts with both computational and experimental approaches.

2. EXPERIMENTS and RESULTS

Impact of BN morphology on BN-polymer composite's thermal performance

h-BN powder made by Momentive can be categorized two major morphologies – platelets and agglomerates. Among agglomerates, there are also agglomerates in irregular shape or close to spherical shape. Various sizes and shapes of platelet BN and agglomerate BN were compounded at specified level (wt. or vol.%) in silicone resin with Crosslinker at 3500 rpm for 30 seconds in Speedmixer. The mixture was then cured in compression mold to produce ~0.5-1.3mm thick pads for through-plane thermal conductivity measurement by Netzsch Laser Flash LFA447. Comparison of in-plane and through-plane TC of BN-silicone composite made with 40wt% platelet BN and spherical agglomerate BN is demonstrated in Figure 1. Spherical BN powder grades deliver higher & more isotropic thermal conductivity than platelet BN at the same loading level and compounding conditions.

Effect of compounding conditions on the through-plane thermal conductivity of BN-polymer composites

PTX60 BN powder is a special BN powder with particles in close to spherical shape. This spherical shape agglomerate provides the best isotropic thermal performance as shown in Figure 1. However, the agglomerate of the building platelets could be broken under high shear when aggressive compounding conditions are applied in compounding process with polymers, which leads to the particle size reduction and the loss of spherical agglomeration. As shown in Figure 2, when compounding with silicone by speed mixer, different combinations of blending speed and time lead to different levels of agglomerate broken down. The morphology change results in significant impact on final BN-silicone composite through-plane thermal conductivity. The final through-plane TC of the same formulation could vary 4x between the lowest and highest values depending on the combination of blending RPM and blending duration. These results demonstrated the importance of choosing suitable compounding conditions when processing agglomerate BN with Polymers.

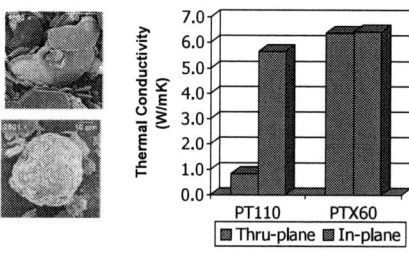

Figure 1. Platelet Vs Spherical agglomerate. PTX60, PTX25, PT371, PT350 are agglomerate BN with different sizes and morphologies, while PT110 and PT120 are platelet BN. Note: Test data. Actual results may vary.

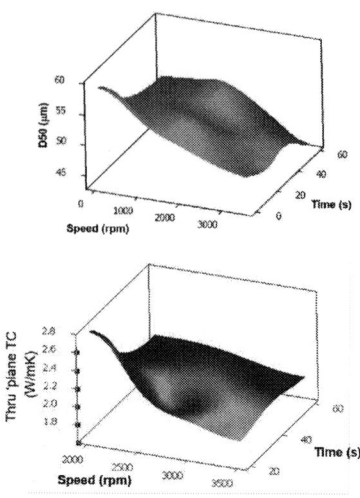

Figure 2. Effect of mixing Speed & Time on BN agglomerate morphology and through-plane thermal conductivity. Note: Test data. Actual results may vary.

TCPs heat sinks Vs Aluminum heat sinks by Finite element analysis

Finite element analysis is used to analyze heat transfer from a heat sink geometry and compare the heat dissipation effects between the heat sinks made of TCPs and anodized aluminum. The model was developed in SOLIDWORKS[†] 14 software and imported into ANSYS[†] 15 software using an IGES format. The model was meshed using the in-built meshing algorithm in ANSYS software to generate a medium mesh.

The heat transfer problem was set up as shown in Table 1 below. A heat flux of 3W was applied on the top surface of the heat sink and a heat transfer coefficient was chosen to reflect convection-limited conditions.

Parameter	Value
Heat sink material	
• Anodized aluminum	100 W/mK
• TCP	3 W/mK
Convection coefficient	5 W/m²K
Emissivity	
• Aluminum	0.77
• TCP	0.9
Ambient temperature	25 °C

Table 1. Parameters used for finite element analysis.

The steady state solution for an aluminum heat sink is shown below in Figure 3 (a). As we can see from the results, the hottest operating temperature was about 69 °C for this case.

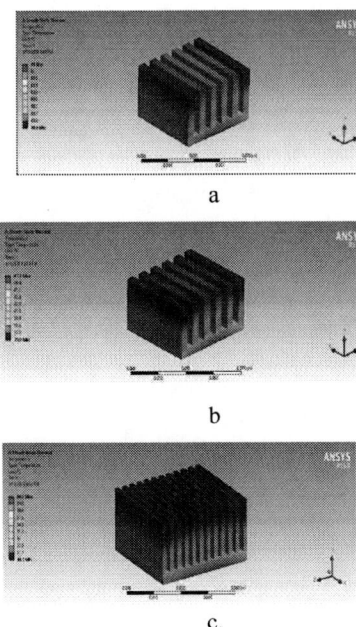

a

b

c.

Figure 3: (a). Steady state solution for anodized aluminum heat sink (6 fins, natural convection and radiation emissivity = 0.77, T max = 41C); (b). Steady state solution for a 3 W/mK TCP heat sink (6 fins, natural convection and radiation emissivity = 0.9, T max = 47.7C); (c). Solution with modified design for 3 W/mK TCP (12 fins, natural convection and radiation emissivity = 0.9, T max = 40.9C). Note: Test data. Actual results may vary.

The maximum operating temperature shown in the above figure essentially represents the operating temperature of the heat source. It is worth noting that this analysis is independent of the nature of the heat source, i.e. the source could be essentially any electronic component like an LED chip an integrated circuit (IC), or an application-specific IC.

Figure 3(b) shows a simplified solution with a 3 W/mK thermally conductive plastic heat sink. The results show that the maximum operating temperature was about 47.7 °C. The results also show that despite the large difference in thermal conductivities (100 W/mK for aluminum versus 3 W/mK for the TCP), there was only a small penalty in the operating temperature of about 6.7 °C.

However, it is important to note that such a direct substitution would be unusual in a metal replacement effort. It would be prudent to take advantage of the design freedom of plastics to improve the design of the heat sink. Since it is possible to injection mold finer features with plastics (as compared to die-cast aluminum), it is not difficult to increase the heat sink surface area with finer fins. The TCP heat sink was redesigned and the solution for the modified 3 W/mK TCP heat sink is shown below in Figure 3(c).

As we can see, with some design modifications, a 3 W/mK TCP may be able to perform comparatively to a far more conductive aluminum heat sink. This verifies the results from the theoretical analysis in the preceding section that in convection limited environments, it may indeed be possible to substitute anodized aluminum with a TCP of modest conductivity and still achieve the desired thermal management solution. Further, TCPs offer the advantages of design freedom, easier parts integration and assembly and

significant weight savings (up to 50%) compared to metal solutions.

3. CONCLUSIONS

High performance thermally conductive, light weight and low coefficient of thermal expansion materials are desired in innovative thermal management designs for dissipating the heat more efficiently in today's high power density electronic devices. Hexagonal boron nitride (h-BN) is unique synthetic high performance thermally conductive ceramic materials with very low density and excellent electrical insulation. Enhanced efficiency in critical heat removal and weight reduction can be achieved simultaneously for various components containing h-BN if right processing conditions are implemented. Theoretical and computational approaches discussed in this study clearly demonstrate the feasibility of replacing aluminum heat sinks with thermally conductive plastic heat sinks in convection-limited environments.

4. REFERENCES

[1] Z. Carl, "Advances in composite materials for thermal management in electronic packaging', JOM, June 1998, Volume 50, Issue 6, pp 47–51.

[2] L. James, J. Peter, "Thermally conductive interface materials and methods of using the same", U.S. Patent No. 5,213,868 A, issued May 25, 1993.

[†]SOLIDWORKS is a trademark of Dassault Systemes Solidworks Corporation. ANSYS is a trademark of ANSYS Inc.

Performance of Durable High-Performance Polymer Composite TIMs Under Accelerated Aging Conditions

Hyungyung Jo and John Howarter

Materials Engineering, Purdue University, 701 W Stadium Avenue, West Lafayette, IN 47907, USA

PH: 765.237.2140 (HJ); 765.496.3103 (JH) E-mail: jo24@purdue.edu (HJ); howarter@purdue.edu (JH)

SUMMARY

Polymer-based thermal interface materials (TIMs), filled with thermally conductive particles for improving thermal performance, are commonly used in electronic devices. A polymer composite TIM would have the desirable attribute of maintaining thermal properties without compromising mechanical compliance and reliability, but thin-film delamination becomes a major concern for composite TIMs as the portion of filler increases. In order to design the desired polymer composite structure, it is vital to understand the adhesion between devices and the film. Various physical properties can influence the failure to adhere, such as bond-line thickness and the Young's modulus of the materials. In electronic packaging systems, the mismatch in the coefficient of thermal expansion (CTE) is also considered to be a main cause of delamination. Because the CTE value of polymer is over 10 times greater than that of metallic fillers, it is crucial to study thermal effects when designing a polymer composite TIM.

The contributions of this paper:

1) This study emphasizes the balance between mechanical and thermal performance, and thus we develop experimental measurements to evaluate mechanical and thermal performance.

2) Model TIMs comprised of polydimethylsiloxane (PDMS) and copper nanoparticles or PDMS and microporous aluminum foam were studied. The aluminum foam was used as a model for a fully percolated interpenetrating network. Various PDMS samples having diverse ratios of crosslinker were investigated in order to study the effects of the bond-line thickness and modulus.

3) The interface adhesion force was measured through the 90-degree peel test and the lap-shear test. We also discuss the thermal performance of highly dense particle-loaded samples by analyzing the microstructure after thermal aging and cycling tests were performed. We compared the mechanical and thermal performances to those of various types of commercial TIMs.

1. INTRODUCTION

Thermal interface materials (TIMs) provide a thermally conductive pathway from heat sources to heat sinks. The materials most commonly used as TIMs in commercial applications are thermal grease and pads[1]. They cannot provide adhesion by themselves, however, and pump-out and phase separation occur frequently. In order to improve thermal performance, various new types of composite material have been introduced[2][3]. Designing polymer-metallic composites is considered to be an approach[4] that would make possible improved thermal performance without compromising mechanical reliability.

During a system's lifecycle, the temperatures of the TIMs, the heat source, and the heat sink all increase and decrease. These elements deform at different rates in this thermal cycle due to their various CTEs[5]. Therefore, the continual thermal cycles can result in a loss of contact between the TIM and the source or sink. The resulting inefficient heat transfer caused by this disconnection has been a well-known and crucial concern in thermal management[6], but many recent studies have not emphasized the balance between mechanical and thermal performance. In this study, we investigate various effects related to mechanical and thermal performance and compare to the data of commercial TIMs.

2. EXPERIMENTAL METHODS

The PDMS and the cross-linking agent were provided by Dow Corning. Copper powders with an average particle size of 40 nm were purchased from US Research Nanomaterials. Various amounts of the cross-linker were applied to produce diverse modulus ranges from 50:1 to 10:1. PDMS composites with 15 Vol% of copper particles were fabricated using a centrifugal mixer. Similar PDMS compositions were used for the infiltrated aluminum foam samples. Glass slides were utilized as substrates to perform initial mechanical and thermal reliability tests for model TIMs which were supported on a single substrate and for sample geometries where the TIM was sandwiched between two glass slides. Data is currently being collected for model TIMs sandwiched between dissimilar substrates of copper and silicon, with the goal of creating realistic CTE mismatches to induce significant mechanical stresses during thermal cycling.

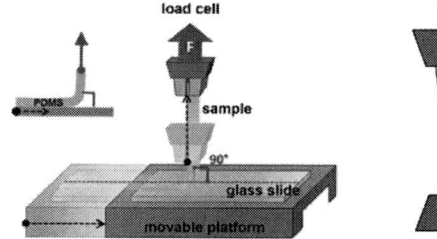

Figure 1. The 90-degree peel test and the lap shear test.

The 90-degree peel test and the lap shear test were performed under perpendicular stress and shear stress, respectively. Both tests were carried out with the same crosshead speed of 0.2 mm/s. The experimental setups are schematically illustrated in Figure 1.

A thermal reliability test was also conducted, using thermal cycling and thermal aging tests. The thermal cycling test was carried out at -40 °C and 130 °C for 5 minutes for each temperature, repeatedly for as few as 5 total thermal cycles. The thermal aging test was performed at 130 °C for 3 days. After the thermal tests, the samples were observed using optical microscopy and scanning electron microscope (SEM) to characterize fracture surfaces and delamination events. Various samples were tested in order to evaluate the effect of film thickness and PDMS modulus on mechanical

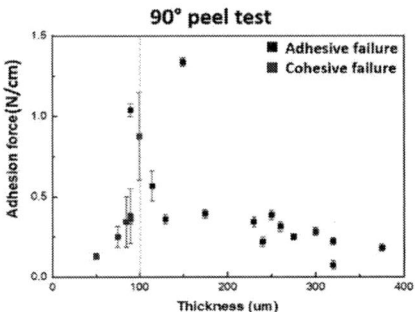
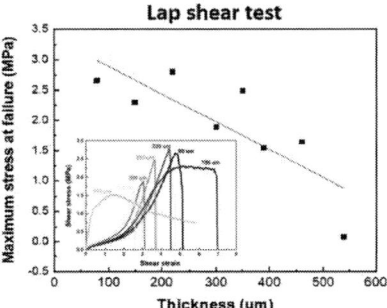

Figure 2. Results of the 90-degree peel test and the lap-shear test.

materials which were handled and processed in the same manner. Thermal cycling likely resulted in the migration of siloxane oligomers previously mentioned and the migration and aggregation of the metallic particles within the polymer matrix. Both mechanisms have the potential to cause areas of localized embrittlement of the TIM resulting in a greater susceptibility for mechanical failure during operation. We intend to compare the performance of copper particle systems with the pre-fabricated metallic foam based TIMs which should not have the challenge of particle migration due to thermal aging or cycling.

and thermal performance. Longer term tests are ongoing which will characterize behavior for model TIMs over 1000 hours of thermal aging and up to 100 thermal cycles.

3. RESULTS

The results of the 90-degree peel test and lap-shear test for the 10:1 PDMS are shown in Figure 2. Generally, a thinner bond-line thickness resulted in a greater adhesion force in both tests[7][8], as depicted in Figure 2 below. The samples that were thinner than 100 μm cohesively failed when the 90-degree peel test was performed, because the thinner films were not able to withstand the peeling force before delamination occurred.

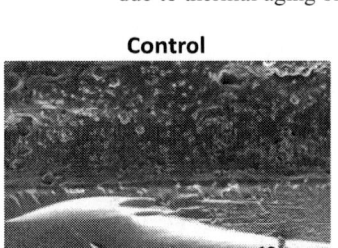

Figure 4. Cross-section SEM images of PDMS + Cu composite before and after thermal cycling.

4. CONCLUSIONS

We develop delamination tests to validate mechanical performance, and we determine thermal and mechanical reliability performance by short-term aging and cycling tests. The final presentation will provide more detailed results including longer-term aging and cycling tests (approximately 1000 hours) and thermal performance measurements for copper and aluminum foam systems. The CTE mismatch will be discussed in order to analyze the failure mechanism with respect to model TIMs bonded between silicon and copper substrates. The mechanical and thermal characterization results will also be compared to the test data of commercial TIMs.

Figure 3. Thermal aging and cycling test results for the sandwich-type sample of PDMS + Cu composite.

Figure 3 shows the results of the preliminary thermal aging and cycling tests for the sandwich-type sample of PDMS + Cu composite. The composite sample was macroscopically observed during the thermal aging and cycling tests. The sample appeared to be drying out, and it delaminated gradually, as shown in Figure 3(a). The drying phenomenon could be due to the loss of silixane oligomers, but there was no significant damage in SEM observation when the cross-sectional fracture surface was analyzed.

While the initial thermal cycling tests did not result in significant mechanical failure or delamination, the cross-sectional analysis of the model PDMS + Cu TIMs showed a dramatic morphological change. After cycling there was significant evidence of internal particle migration and cracking of the TIM. Some cracking may have resulting from the sectioning and sample preparation processing, however, this phenomenon was not observed in the initial

5. REFERENCES

[1] F. Sarvar, D. C. Whalley, and P. P. Conway, "Thermal interface materials - A review of the state of the art," *ESTC 2006 - 1st Electron. Syst. Technol. Conf.*, vol. 2, pp. 1292–1302, 2007.

[2] K. M. F. Shahil and A. A. Balandin, "Graphene-multilayer graphene nanocomposites as highly efficient thermal interface materials," *Nano Lett.*, vol. 12, no. 2, pp. 861–867, 2012.

[3] K. Uetani, S. Ata, S. Tomonoh, T. Yamada, M. Yumura, and K. Hata, "Elastomeric thermal interface materials with high through-plane thermal conductivity from carbon fiber fillers vertically aligned by

electrostatic flocking," *Adv. Mater.*, vol. 26, no. 33, pp. 5857–5862, 2014.

[4] B. Carlberg and D. Shangguan, "Nanostructured polymer-metal composite for thermal interface material applications," *2008 58th Electron. Components Technol. Conf.*, pp. 191–197, 2008.

[5] W. Lin, K. S. Moon, and C. P. Wong, "A combined process of in situ functionalization and microwave treatment to achieve ultrasmall thermal expansion of aligned carbon nanotube-polymer nanocomposites: Toward applications as thermal interface materials," *Adv. Mater.*, vol. 21, no. 23, pp. 2421–2424, 2009.

[6] R. Prasher, "Thermal interface materials: Historical perspective, status, and future directions," *Proc. IEEE*, vol. 94, no. 8, pp. 1571–1586, 2006.

[7] M. D. Banea, L. F. M. Da Silva, and R. D. S. G. Campilho, "The effect of adhesive thickness on the mechanical behavior of a structural polyurethane adhesive," *J. Adhes.*, vol. 91, no. 5, pp. 331–346, 2014.

[8] T. J. W. Wagner and D. Vella, "The 'Sticky Elastica': Delamination blisters beyond small deformations," *Soft Matter*, vol. 9, no. 4, pp. 1025–1030, 2013.

The Impact of Anodization on the Thermal Performance of Passively Cooled Electronic Enclosures Made of Die-cast Aluminum

Zhongchen Zhang [1], Michael Collins [2], Chris Botting [3], Eric Lau [3], Majid Bahrami [1] *

[1] Laboratory for Alternative Energy Conversion, School of Mechatronic Systems Engineering,
Simon Fraser University, Surrey, British Columbia, Canada
[2] Solar Thermal Research Laboratory, Department of Mechanical and Mechatronics Engineering,
University of Waterloo, Waterloo, Ontario, Canada
[3] Delta-Q Technologies, Burnaby, British Columbia, Canada
* Corresponding author – email: mbahrami@sfu.ca

Abstract

Natural convection and thermal radiation from anodized die-cast aluminum enclosures designed for the application of high power density battery chargers are investigated experimentally. Several sample enclosures are prepared using the two most commonly available types of anodization: i) Type II-black; and ii) Type III-clear. The enclosures are then tested in a customized natural convection - thermal radiation test chamber. Total emittance of bare and anodized surface is measured using Fourier Transform Infrared Reflectometer (FTIR) spectroscopy. The results show that the total hemispherical emissivity of die-cast aluminum sample surface can be significantly improved from 0.14 to 0.92 after anodizing. More importantly, anodization can lead to a considerable reduction in thermal resistance (up to 14.7%) compared to the identical untreated enclosures. Because of the unique surface morphology of anodized die-cast aluminum, it is also observed that the same improvement of thermal emissivity as well as overall thermal performance can be achieved by either types of anodization.

Keywords

Thermal radiation, Natural Convection, Anodization, Electronic cooling.

Nomenclature

A Surface area [m^2]
C_p Specific heat at constant pressure [J/(kg · K)]
C_3 Third radiation constant [μm · K]
E Surface radiative emissive power [W/m^2]
F View factor
H Enclosure height [m]
h Convective heat transfer coefficient [W/(m^2 · K)]
I Current [A]
k Thermal conductivity [W/(m · K)]
L Enclosure length [m]
Q Heat transfer rate [W]
R Thermal resistance [°C/W]
s Standard deviation
T Temperature [°C]
V Voltage [V]
W Enclosure width [m]

Greek Symbol
σ Stefan-Boltzmann constant [W/m^2 · K^4]
ε Thermal emissivity

δ Absolute uncertainty
λ Wavelength [μm]

Subscript
b Black body
c Convective heat loss
r Radiative heat loss
h Heat sink
∞ Ambient environment

1. Introduction

Reliable thermal management of electronic devices is always crucial, and of great importance, especially in power electronic industry, where the cooling loads are ever-increasing. Active cooling technologies, on the basis of force convection, are most widely used because of their relatively higher cooling capacity. However, active cooling require parasitic power, can be noisy (e.g., fan, pump) and poses a risk of device overheat and malfunction induced in case of the failure. Passive cooling technologies work based on natural convection and thermal radiation, and offer a reliable alternative cooling solution with no additional moving part and zero parasitic power consumption, which is desirable in electronic cooling applications, when applicable.

In theory, convective heat transfer can be calculated by Newton's Law of cooling as Eq. (1):

$$Q_c = hA(T_h - T_\infty) \qquad (1)$$

Where, h is the convective heat transfer coefficient, A is the heat transfer surface area, and T_h, T_∞ are the average surface and ambient temperature, respectively. Meanwhile, the modified Stefan-Boltzmann equation [1] can be used to calculate radiative heat exchange from a surface to the ambient:

$$Q_r = AF\sigma\varepsilon(T_h^4 - T_\infty^4) \qquad (2)$$

where, F is the view factor from surface to surrounding, σ is the Stefan-Boltzmann constant [5.67×10^{-8} W/m^2·K^4] and ε is the surface thermal emissivity, respectively. Numerous studies were focused on the enhancement of natural convection by investigating different geometries. Studies by Ahmadi et al. [2] [3] are good examples illustrating the natural heat transfer improvement by introducing the interrupted fins. However, radiative heat transfer is overlooked in the majority of the

Table 1. Literature review on the radiative heat transfer from finned enclosures

Ref.	Approaches	Fin Geometry	Material	Surface Treatment	Surface Emissivity	Emissivity Measuring Method	Qr/Q(%)
Chaddock et. al [4]	Experimental	Cylindrical	Extruded Aluminum	-	0.99	-	33
				Polished	-	-	10 - 20
Sparrow et. al [5]	Experimental Analytical	Pin	Extruded Aluminum	Black Anodizing	0.82	Gier-Dunkle heated-cavity reflectometer.	25 - 45
Rao et. al [6]	Experimental Numerical	Rectangular	Extruded Aluminum	Black Board Paint	0.85	-	25 - 40
Rao et. al [7]	Numerical	Rectangular	Extruded Aluminum	-	0.85	-	36 - 50
Yu et. al [8]	Experimental Analytical	Rectangular	Pure Aluminum	Black Anodizing	0.80	-	27
Tamayol et. al [9]	Experimental Analytical	Rectangular	Extruded Aluminum	-	0.75	-	50

existing studies and is given less consideration for manufacturing the naturally-cooled heat sinks. Nevertheless, more evidence is emerging that radiation play a critical role in passively cooled devices. One of the pioneering studies on this topic was conducted by Edwards et al. [4]. They reported that the contribution of radiation heat transfer was one third of the total heat dissipation from an extended cylindrical surface. Following studies further affirm that great portions of heat loss in passive cooled devices are contributed by radiation regardless of its external geometry [4] to [9]. Table 1 provides an overview of the pertinent literatures that support the important role of thermal radiation in the passive cooled systems.

In radiative heat transfer, surface area, view factor from enclosure to ambient, and surface thermal emissivity are the three major parameters when the temperature difference is fixed, as it can be seen in Eq. (2). The surface area and view factor are all associated with heat sink geometry, which are often limited by the actual product design. The surface emissivity, as a surface radiative property, often can be improved by several surface finishing techniques. Thus, this study focuses on the heat transfer enhancement enabled by the thermal emissivity improvement.

The majority of heat sinks used in the industry are made of aluminum alloys, either from extrusion, casting, or machining process. Several methods are often used to improve the thermal emissivity of aluminum surfaces, such as abrasive blasting, spray painting, anodizing. Of all the techniques, anodizing is the most efficient and cost-effective method. Known as a process of growing an anodic layer integrated with the surface of aluminum alloys in an electrolytic bath, it is usually adopted to enhance aesthetic appeal, along with wear and corrosion resistance to the original part in many applications. According to the US Military specification [MIL-A-8625F] [10], anodic coating for aluminum and its alloys are categorized into 3 types and 2 classes. Due to the environmental concerns, Type I [Chromic acid anodizing] is only used in rare cases while Type II [Sulfuric acid anodizing] and Type III [Hard anodic coatings] are the mostly common and well-adopted techniques. Attribute to the porous structure of the anodic layer, the dyes and pigments can be readily absorbed and sealed in the small pores and result in various colorful appearance, also referred as class I, or colored anodizing. Subsequently, class II is the clear anodizing with original look of the anodic layer itself.

In general, most of previous studies either solely emphasized on the experimental and modeling work of thermal radiation in naturally cooled heat sinks as listed in Table 1 or investigated the influence of the key parameters in anodizing process. These parameters have an important impact on the morphology of the anodic layer, which will in turn affect the total emittance of treated surface as shown in [11] [12] [13]. To the best of the authors' knowledge, conjugate analysis involving both aspects has not been investigated thoroughly in the literature. We also noticed that no study has placed their focus on the thermal emissivity of anodized die-cast aluminum. This could be due to the limited availability of the material. Therefore, in this study, die-cast aluminum battery charger enclosures [Delta-Q Technologies] are anodized in Type II-black and Type III-clear, by a local metal surface treatment company [Spectral Finishing]. The overall thermal performance of three types of heat sink with different surface finishes is tested and analyzed. The thermal emissivity of the tested heat sinks is also measured and reported.

2. Experimental Approach

2.1 Sample Preparation

Several unprocessed enclosures were prepared and cleaned in advance for anodizing and testing, as shown in Fig. 1. All samples were made from die-cast process using aluminum A380 alloy. The length [L], width [W] and height [H] of the enclosures were 23cm, 17cm and 7.5cm, respectively. Isopropyl alcohol solution was used to remove dusts and stains from the enclosure surface before anodizing.

This electrochemical reaction occurred in a 15% sulfuric acid bath with special additives at 14°C to achieve a 25μm thick anodic layer for both Type II and Type III. Because of the high content of impurities in this type of die-cast aluminum alloys, increasing the thickness of anodic layer any further is of great difficulty. In Type II-Black anodizing, the black dyes were added into the electrolyte bath to reach the desired dark black finish that absorbed through capillary effect. Finally, the cold nickel fluoride method was applied to accomplish proper sealing. Two sample enclosures were prepared for each type of anodization to ensure repeatability of the testing results. The detailed numbers of sample preparation are provided in Table 2.

Table 2. Number of samples prepared with various surface finish

Amount of samples	Bare	Type-II Black	Type-III Clear	Type-III no sealing
Enclosures	2	2	2	-
Plates	2	2	2	2

Additionally, eight flat plate samples [2.5cm × 5.0cm] were prepared with the same material, cleaned and anodized with the same procedures for thermal emissivity measurements.

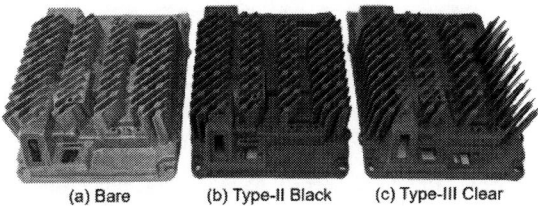

(a) Bare   (b) Type-II Black   (c) Type-III Clear

Figure 1. Delta-Q battery charger enclosures, samples made using die-cast process, aluminum A380 alloy; (a) before and after various types of anodization (b) and (c).

2.2 Test Procedure

Figure 2 shows the schematic of our experimental setup. A customized natural convection - thermal radiation test chamber was built with 4 mm thick acrylic plastic [k = 0.2W/(m · K), $\rho = 1180$kg/m^3, $C_p = 1470$J/(kg · K), $\varepsilon = 0.9$]. The length, width, and height of the chamber were 50 cm [2.2 x L], 50 cm [3.0 x W] and 50 cm [6.6 x H] of the tested enclosure to minimize the influence of chamber walls on the internal natural convection airflow. This chamber was built to reduce the impact of the air disturbance from surroundings as well as to improve the result consistency from various tests.

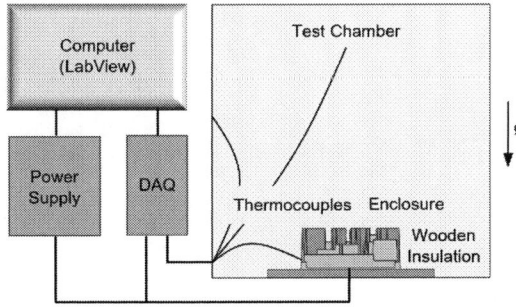

Figure. 2: Schematic diagram of experimental setup.

The sample battery charger enclosures were tested in the horizontal orientation. A centimeter-thick wooden plate was placed underneath the enclosure to provide thermal insulation from the base. The ports and gaps on the enclosure were taped with adhesive aluminum foils to ensure airtight sealing, preventing the potential thermal leakage by air convection from the internal regions. Five ultra-thin polyimide film heaters with pressure sensitive adhesive of three different sizes of 1.3cm×5.1cm [0.5in×2.0in], 2.5cm×2.5cm [1.0in×1.0in] and 2.5cm×5.1cm [1.0in×2.0in] were attached to the back of the enclosure to mimic the heat generation from diodes, transistors, inductors, transformers and other heat generating components.

All heaters had a power density of 1.6W/cm^2 [10W/in^2] and were driven by a programmable AC/DC power supply. Fourteen T-type copper-constantan thermocouples with uncertainty of ±0.5°C were installed using aluminum foil tapes at various positions to monitor the temperature distribution along the enclosure base, chamber inner ambient and chamber wall. The location of the thermocouples and heating components is indicated in Fig.3. A NI9213 [National Instruments] data acquisition module was used to record the temperature readings from the thermocouples and a NI9229 [National Instruments] for monitoring the voltage and current through the heater to calculate the actual supplied power.

The tests were run in an open lab environment facing north, free of direct sunlight from the windows. The room temperature was kept constant at 22°C. Each enclosure was tested with various power levels ranging from 20W to 80W. Steady-state condition was reached when the partial derivative of all temperatures with respect to time - except the ambient - was within 0.001°C for 30 mins. This was considered as the thermal equilibrium on the interfaces of the tested enclosures and chamber walls, i.e., the summation of convective and radiative heat transfer leaving each interface became equal to the heat input into the battery charger.

The thermal emissivity of each anodized surface was measured using Fourier Transform Infrared Reflectometer [FTIR] spectroscopy [400T, Surface Optics Corporation] at room temperature, where the measured range of the wavelength of infrared is from 2.5µm to 25µm. The spectral hemispherical emissivity and total hemispherical emissivity of sample plates were determined in the same ambient condition. The micro structures of anodic layer were imaged by a field mission scanning electron microscopy [Nova NanoSEM, Thermo Fisher Scientific]. All samples were coated with 10nm iridium using a high vacuum sputter system [EM ACE 600, Leica Microsystems] before SEM imaging.

3. Results and discussions

3.1 Thermal emissivity of anodized surface

Emissivity as a surface radiative property specifies the radiative capability of a real surface as compared with emission from a black body at the same temperature. In general, it depends on two factors, the temperature and direction. The emissivity value - that is widely accepted and reported in most textbooks - is the total hemispherical emissivity. This value should be averaged over all the wavelengths and the directions. However, most data are only available for a limited range of wavelength and directions due to the constrains of measurement and the interests of research, i.e., emissivity over mid- and long-infrared [MWIR, LWIR] is important in thermal applications. Additionally, it can also be spectral. The Wien's Displacement Law [1] expresses the spectral distribution of a black body with the corresponding wavelength to temperature as Eq. (3):

$$\lambda_{max} \cdot T = C_3 \tag{3}$$

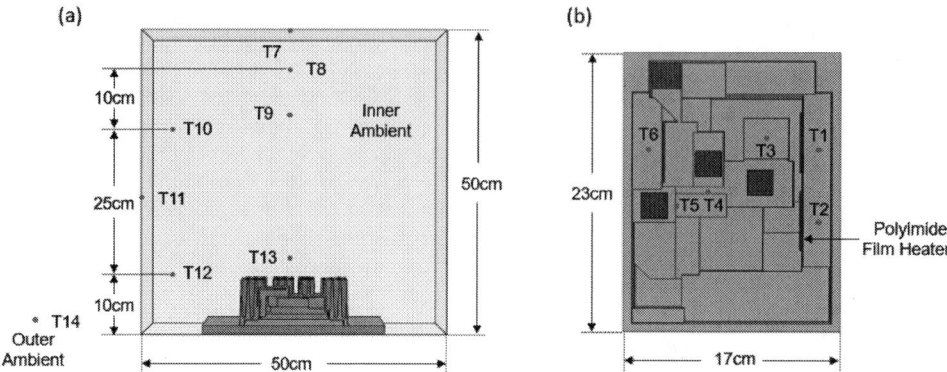

Figure 3. Schematic of thermocouples locations and heat generating components in (a) test chamber (b) battery charger enclosure.

where, the λ_{max} is the peak wavelength and C_3 is the third radiation constant, $2898\mu m \cdot K$, respectively. According to Eq. (3), the maximum spectral emissive power is shifting towards shorter wavelength as temperature increases. For a blackbody at 100°C [373.16K], the peak emission occurs over 7.8μm and great portion of the emitted radiation fall into the range of mid-infrared [MWIR]. It further ascertains that most of the radiation from electronic cooling applications could be found within this range. Ideally, it would be preferable to keep the sample at certain temperature to simulate the real working environment during the measurement. However, conditioning the ambient environment of the spectrometer is challenging.

Figure 4 (a) shows the spectral hemispherical emissivity of various samples at wavelengths from 2.5μm to 25μm. The spectral emissivity justifies the radiation ability of measured surface to the black body at each wavelength at the same temperature, and shows as Eq. (4):

$$\varepsilon_\lambda(\lambda, T) = \frac{E_\lambda(\lambda, T)}{E_{\lambda b}(\lambda, T)} \qquad (4)$$

where, E is emissive power and subscript λ, b denote spectral emissive power from measured surface and a black body, respectively. Surprisingly, the spectral emissivity of a bare die-cast aluminum surface is very selective and has the rising trend towards the short wavelength. It can reach 0.4 at wavelength of 2.5μm while drops to 0.1 at 25μm. This varying properties of bare die-cast aluminum may imply the potential enhancement of thermal radiation in operational scenarios with high temperature. As for anodized surfaces, the emissivity over the entire measured spectrums fluctuates within the range of 0.85 to 1. This significant improvement of emissivity can be gained by either types of anodization. Figure 4(b) shows the results for total hemispherical emissivity which are the quotient between the integration of spectral emissivity over the wavelength of 2.5μm to 25μm and the total emissive power of a black body at the same temperature. The following formulation [1] can be used to calculate total emissivity:

$$\varepsilon(T) = \frac{\int_{\lambda=2.5}^{\lambda=25} \varepsilon_\lambda(\lambda, T)E_{\lambda b}(\lambda, T)d\lambda}{\sigma T^4} \qquad (5)$$

The error bars shown in Fig. 4(b) are comprised of the inaccuracy of the spectrometer [1%] and the standard deviation between different samples. The following can be observed from Fig. 4(b):

- The total emissivity of the type II-Black treated surface can reach 0.92 and it is improved nearly seven times compared to its original surface condition;
- The total emittance of both type II-Black and type III-Clear are nearly identical. This indicate that the black dyes may have limited contribution in terms of thermal emissivity improvement and may imply that the color of anodized surface not certainly reflect its own thermal emissive capability.

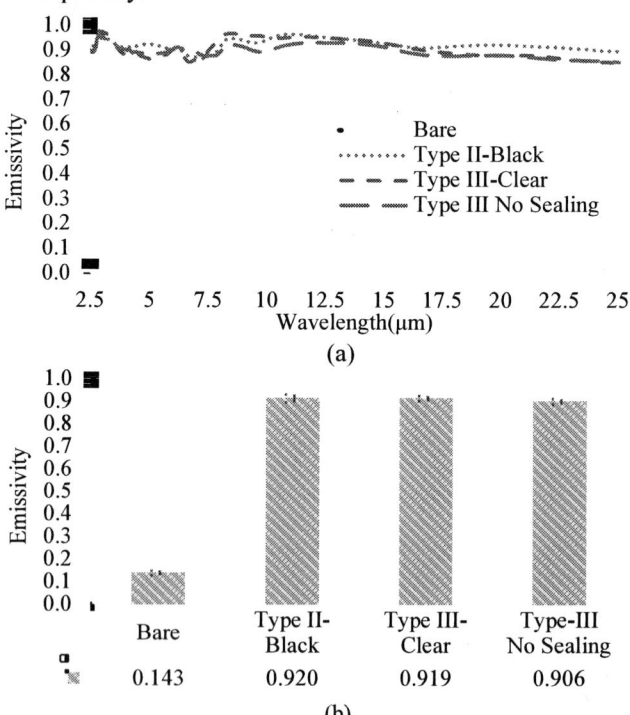

Figure. 4: (a) Spectral Hemispherical Emissivity (top) and (b) Total Hemispherical Emissivity (bottom) of Bare, anodized Type II-Black, Type III-Clear and no sealing sample.

In terms of the whole spectrum of light, the wavelength of electromagnetic radiation in the range of 390nm to 700nm is classified as visible light that is involved in the perception of different color in human eyes. Meanwhile, the infrared spectrum of thermal radiation is in the wavelength of 3μm to 15μm, the Mid-wavelength infrared to Long-wavelength infrared, which can be found usually within the application of power electronics and is of great interest in thermal radiation. It is noticeable 10x order of magnitude difference in the wavelength of spectrum distribution of visible light and infrared light that corresponding to the thermal radiation, which helps explaining the fact that colored surface offers the same thermal emissivity as un-dyed surface and the blackness of the surface is not necessarily associated to its emissive property within infrared region. It also reveals a possibility to maintain the same thermal improvement using type II anodization with colorful dyes to meet the various marketing needs. However, we still need to consider the thermal radiative properties of the dyes and its possible impact on the overall thermal emissivity, which require further investigation.

Figure 4 shows the influence of sealing on the thermal emissivity as well. Sealing is a process of precipitating an additional layer of sealants, seeing as a thin film, on top of the uncovered anodized surface to protect the pore structure and the dyes that are already absorbed inside. The easiest sealing method is called hydrothermal, which is applied by placing the parts in a hot water bath. Several other methods are also used to fulfill the same purposes. Among all, the cold nickel fluoride method is chosen by this anodizing vendor and it does help the total emittance increase slightly, i.e. from 0.906 to 0.919. Lee et al [12] have examined the effect of several sealing methods on the surface emissivity and concluded that the improvement after sealing, may attribute to the nature of the precipitated sealants in each method.

3.2 Surface morphology of anodic layers

Figure 5 shows the SEM images of the unsealed anodic layer which unveil the porous structure under the sealed film. The anodic layer consists of cells, in which each has a distinct boundary that separates it from others, shown in Fig. 5 (a). In addition to the unique hexagonally packed cells, Fig. 5 (c) displays a rather different surface structure, where the anodic films appear to be stacked layer-by-layer with small gaps and a few miniature pores scattering over the surface. With higher image magnification in Fig.5 (b) and (d), the details of two totally different morphologies provide a potential explanation for this phenomenon, namely, the density difference within the same die-cast aluminum sample. This commonly exists in the die-cast aluminum since the injection pressure of the molten metal may vary at different positions during the casting process. As such, the density of aluminum sample is unevenly distributed and porous textures are formed resulting in a layer-stacked structure resembling Fig. 4(c) and (d).

Unlike the porous anodic layer that formed on the surface of extruded aluminum alloys [12], the anodized surface of die-cast aluminum does not contain any individual pores aligned within the cells. The certain pattern of pores should be visible at higher magnification, but we fail to observe any arrangement that confirm its existence. Even with image magnification of 46,266×

in Fig. 5(e), the surface structure remains as pore-less as stacked structures. This is mainly due to the composition of the aluminum alloys that is used in die-cast manufacturing, where the pure aluminum only takes up 80% to 90%. The rest of the constituents are mainly silicon, copper, zinc, other metallic elements and impurities from the manufacturing process. It is assumed that the existence of aforementioned contents interrupts the formation and growth of the anodic layer.

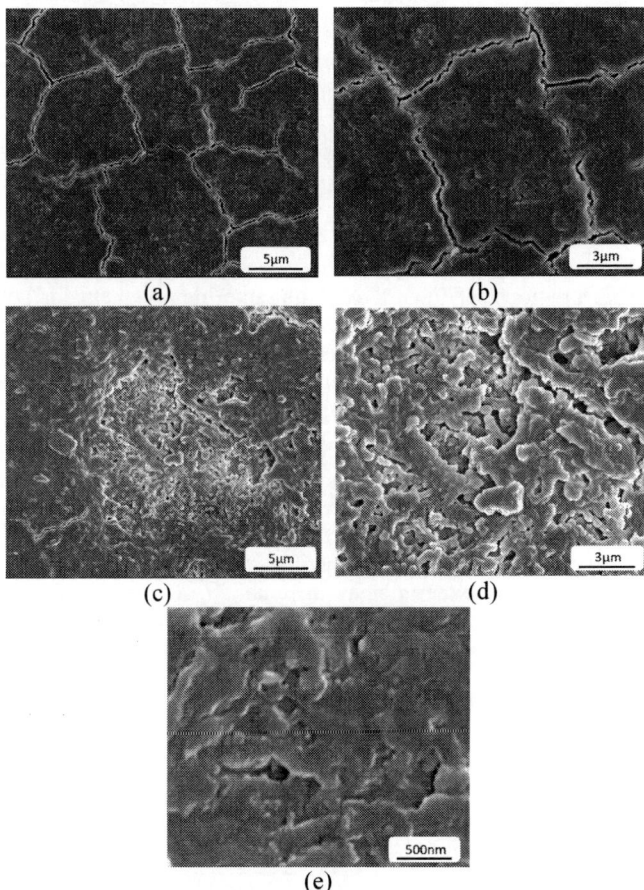

Figure. 5 SEM images of unsealed anodic layers of anodized die-cast aluminum. (a) and (b) show "hexagonally packed" cells; (c) and (d) indicate the porous and layer-stacked structure; (e) porous structure with higher magnification.

Figure 6 shows the surface morphology of sealed anodized surface after type II-Black and type III-Clear treatment. The boundaries or cracks that shape the cells are still partially visible from both images, which may be because of the relatively thin film from the sealing process. However, from the perspective of surface uniformity, the type III-Clear treated surface has comparatively high surface roughness compared to type II-black. This may lead to a potentially higher resistance to wear and corrosion. As a result of the pore-less and layer-stacked structure of anodized die-cast surface, the black dyes may have absorbed insufficiently within the anodic layer in type II-Black treatment as show in Fig. 6(a) as small stains that on the surface.

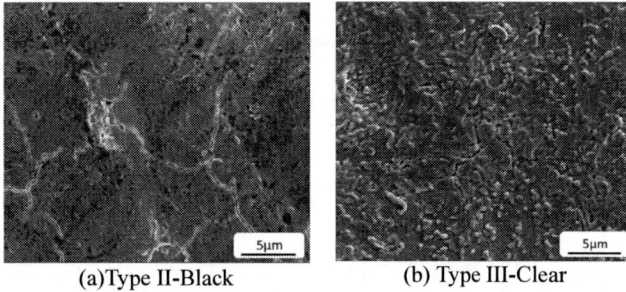

(a)Type II-Black   (b) Type III-Clear

Figure. 6 SEM images of sealed anodic layer of (a) Type II-Black and (b) Type III-Clear. (a) the surface of type II-Black under scanning electron microscope where we assume most of dyes are rather visible and attached on the surface after sealing. It looks like stains on the surface. (b) shows the morphology of type III-Clear treatment and it appears to be a relatively rough surface.

Even though the micro surface morphology appears to be different after the sealing process, the emissivity measurement from previous sections reveals the same improvement for both types of anodization. Consequently, we may able to conclude that, the unique surface structure made by anodizing process, i.e. porous layer, has a fundamental impact on thermal emissivity, and that black dyes and sealing process are less important.

3.3 Thermal performance of anodized enclosure

The thermal test results are shown in Fig. 7, which are the average temperature difference between the enclosure base and the ambient temperature, and the total thermal resistance of the enclosure, including both natural convection and thermal radiation. The experimental uncertainty of temperature and power input measurements as well as standard deviation among each test and sample are shown as error bars in Fig.7. The horizontal error bars represent the calculated uncertainty from voltage and current measurements while the vertical uncertainty includes the calculated errors of temperature difference measurements and standard deviation from tests for the same sample. Since the maximum calculated uncertainty for temperature measurement is $\pm 4.7\%$, some error bars in Fig. 7 (a) are rather invisible. Uncertainty analysis are listed below.

$$\delta_{\Delta T} = \sqrt{(\delta_T)^2 + (\delta_T)^2} \qquad (6)$$

$$\delta_{\Delta T \cdot s} = \sqrt{(\delta_{\Delta T})^2 + s^2} \qquad (7)$$

$$\frac{\delta_{\dot{Q}}}{\dot{Q}} = \sqrt{\left(\frac{\delta_V}{V}\right)^2 + \left(\frac{\delta_I}{I}\right)^2} \qquad (8)$$

$$\frac{\delta_R}{R} = \sqrt{\left(\frac{\delta_{\Delta T}}{\Delta T}\right)^2 + \left(\frac{\delta_{\dot{Q}}}{\dot{Q}}\right)^2} \qquad (9)$$

$$\delta_{R \cdot s} = \sqrt{(\delta_R)^2 + s^2} \qquad (10)$$

As shown in Fig. 7, anodization significantly improves the overall thermal performance of treated enclosures compared to the bare samples. In Fig. 7(a), the average base temperature drops around 6°C on average, 12.2% in relative improvement, for an input thermal power of 80W. The average temperature of the base is also 3°C lower [16.7% improved] even for a power input of 20W. The radiative heat transfer follows the surface

temperature to the 4th power, Eq. (2). Indeed, for higher surface temperatures, the significance of thermal emissivity (and thermal radiation) will be more pronounced.

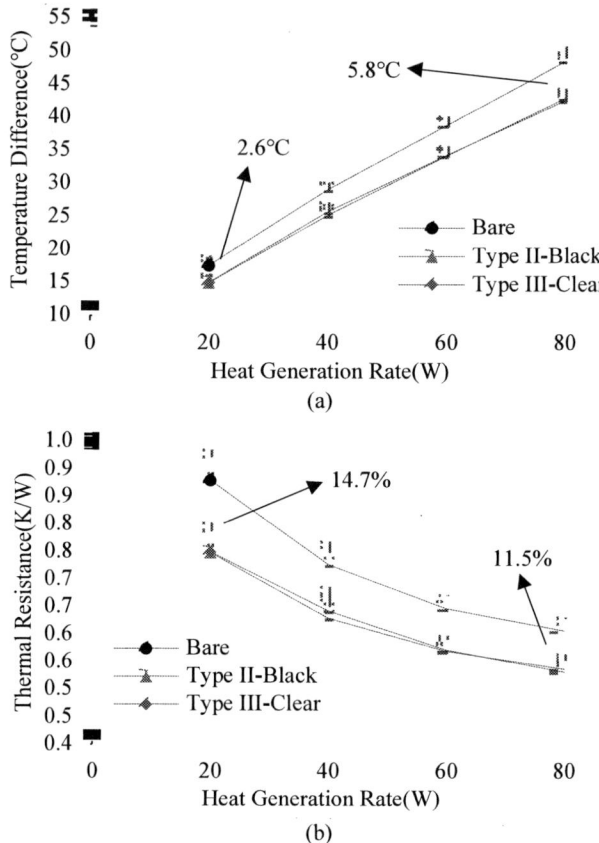

Figure. 7 (a) Average temperature difference between enclosure base and chamber ambient (b) Total thermal resistance (natural convection and thermal radiation) from tested enclosures.

The total thermal resistance of the enclosure can be improved up to 14.7% when after applying Type II-Black anodizing. The same enhancement was observed for Type III-Clear in our tests. This further confirms the significant impact of anodization on the overall thermal enhancement of die-cast aluminum enclosures.

4. Conclusions

A comprehensive experimental study was performed to investigate the impact of anodization on thermal radiation of naturally-cooled heat sinks. A number of die-cast aluminum enclosures were prepared and tested in our custom-designed testbed. The results indicated that the total emissivity of samples can be improved notably from 0.14 to 0.92 after using commercially-available anodizing treatments.

It was also shown that anodized die-cast aluminum naturally-cooled heat sinks perform significantly better, up to 15% reduction in overall thermal resistance, compared to the identical non-anodized enclosures. Due to the unique surface morphology of anodized die-cast aluminum, both types of anodization, Type II-Black and Type III-Clear, were able to provide the same enhancement in thermal radiation.

Acknowledgments

This research is supported by the funding from the Natural Sciences and Engineering Research Council of Canada (NSERC) Collaborative Research Development (Grant No. CRDPJ488777) and Delta-Q technologies. The author also would like to thank the 4D Labs of Simon Fraser University for their support.

References

[1]. Bergman, T. L., et al. "Fundamentals of heat and mass transfer", John Wiley & Sons Inc, USA (2011).

[2]. Ahmadi, M., Mostafavi, G. and Bahrami, M., "Natural convection from rectangular interrupted fins", International Journal of Thermal Sciences, 82, pp.62-71, 2014.

[3]. Ahmadi, M., Pakdaman, M.F. and Bahrami, M., "Pushing the limits of vertical naturally-cooled heatsinks", Calculations and design methodology, International Journal of Heat and Mass Transfer, 87, pp.11-23, 2015.

[4]. Edwards, J.A. and Chaddock, J.B., "An experimental investigation of the radiation and free convection heat transfer from a cylindrical disk extended surface", Trans. Am. Soc. Heat. Refrig. Air-Cond. Eng, 69, pp.313-322, 1963.

[5]. Sparrow, E.M. and Vemuri, S.B., "Natural convection /radiation heat transfer from highly populated pin fin arrays", Journal of heat transfer, 107(1), pp.190-197, 1985.

[6]. Rao, V.R. and Venkateshan, S.P., "Experimental study of free convection and radiation in horizontal fin arrays", International Journal of Heat and Mass Transfer, 39(4), pp.779-789, 1996.

[7]. Rao, V.D., Naidu, S.V., Rao, B.G. and Sharma, K.V., "Heat transfer from a horizontal fin array by natural convection and radiation—A conjugate analysis", International journal of heat and mass transfer, 49(19-20), pp.3379-3391, 2006.

[8]. Yu, S.H., Jang, D. and Lee, K.S., "Effect of radiation in a radial heat sink under natural convection", International Journal of Heat and Mass Transfer, 55(1-3), pp.505-509, 2012.

[9]. Tamayol, A., McGregor, F., Demian, E., Trandafir, E., Bowler, P., Rada, P. and Bahrami, M., January. "Assessment of thermal performance of electronic enclosures with rectangular fins: a passive thermal solution", In ASME 2011 Pacific Rim Technical Conference and Exhibition on Packaging and Integration of Electronic and Photonic Systems (pp. 269-276). American Society of Mechanical Engineers, 2011.

[10]. United Stated Military Specification No. MIL-A-8625F, 1993.

[11]. Le, H.G., Watcher, J.M. and Smith, C.A., "Comparison of sulfuric and oxalic acid anodizing for preparation of thermal control coatings for spacecraft", 1988

[12]. Lee, J., Kim, D., Choi, C.H. and Chung, W., "Nanoporous anodic alumina oxide layer and its sealing for the enhancement of radiative heat dissipation of aluminum alloy", Nano energy, 31, pp.504-513, 2017.

[13]. Kumar, C.S., Sharma, A.K., Mahendra, K.N. and Mayanna, S.M., "Studies on anodic oxide coating with low absorptance and high emittance on aluminum alloy 2024", Solar energy materials and solar cells, 60(1), pp.51-57, 2000.

[14]. Klampfl, B.F., "A parametric study of sulfuric acid anodized 5657 aluminum alloy coatings for thermal control applications", Doctoral dissertation, Rice University, 1998.

Development of a 3D Printed Loop Heat Pipe

Bradley Richard, William G. Anderson, Joel Crawmer
Advanced Cooling Technologies, Inc.
1046 New Holland Ave.
Lancaster, PA USA
bradley.richard@1-act.com

Abstract

CubeSats and Smallsats have increased in popularity and capability, but advanced thermal solutions are required to keep up with the increasing heat loads. Loop heat pipes (LHPs) are a passive solution capable of transporting heat from electronics to deployable radiator panels, but are currently cost prohibitive for most small satellite applications. A 3D printed LHP has been developed using a direct metal laser sintering (DMLS) process. The additive manufacturing process significantly reduces fabrication costs by eliminating labor intensive steps while also eliminating the knife-edge seal to offer improved reliability. The 3D printed wick for the LHP evaporator was developed through an experimental optimization study. A proof of concept LHP was built and tested using a 3D printed primary wick. A maximum power of 125W was achieved during steady state testing. Additional tests were completed including power cycles, adverse elevation, and low power startup. Based on the successful prototype testing additional work is underway for further optimization of the 3D printed wick design and development of DMLS parameters for secondary wick fabrication.

Keywords

Loop Heat Pipe, Additive Manufacturing, Wick, CubeSats, Thermal Management

Nomenclature

P_b : Bubble point pressure (Pa)
R_p : Pore radius (m)
σ : Surface tension (N/m)

1. Introduction

CubeSats and SmallSats have become widely used due to their low costs, fast development times, and increases in capabilities with advanced electronics and deployable mechanisms. The increase in capabilities for compact satellites, however, also results in higher heat loads which require an advanced thermal management solution beyond the use of high thermal conductivity materials for heat spreading. Loop heat pipes (LHPs) are a passive solution which are capable of transporting heat from electronics to deployable radiators, but they are currently cost prohibitive for most CubeSat and SmallSat applications. The use of a 3D printed wick has the potential to significantly reduce LHP fabrication costs while also improving long term reliability. The use of a DMLS fabrication process can eliminate multiple labor-intensive fabrication steps including machining vapor grooves in the primary wick and wick insertion into an aluminum saddle. The ability to print solid and porous wick material together in a single part eliminates the need for the conventional knife-edge seal which improves reliability.

LHPs are passive devices capable of high thermal conductances across long distances. A schematic of an LHP is provided in Figure 1. Heat enters the evaporator and vaporizes the working fluid. The vapor passes through grooves in the primary wick and through the vapor line to the condenser. Here the vapor is condensed and subcooled. The subcooled liquid passes through the bayonet tube into the center of the primary wick. A secondary wick allows for communication between the compensation chamber and center of the primary wick. The compensation chamber contains saturated fluid at a lower pressure than the evaporator which provides the driving force for fluid flow. The capillary pressure of the primary wick must be greater than the total pressure drop of the system to pump liquid from the liquid line return to the evaporation site at the vapor grooves. [1]

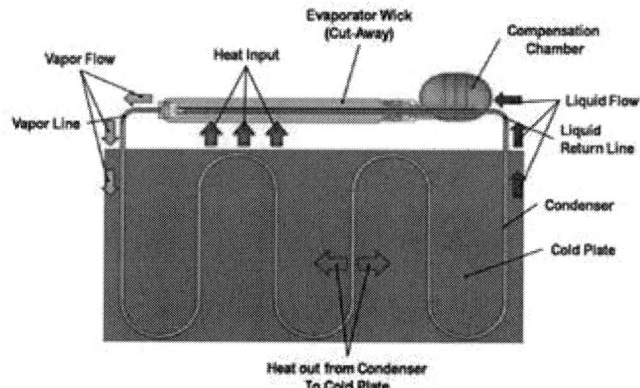

Figure 1: Loop heat pipe schematic (not to scale).

The performance of an LHP primary wick depends largely on pore size. The capillary pressure of a wick is inversely proportional to pore radius. For successful LHP operation the capillary pressure must be greater than the total system pressure drop. Therefore, by reducing the pore size of the primary wick the maximum power can be increased. Traditionally sintered primary wicks have a pore radius of approximately 1μm. DMLS has been demonstrated to be able of printing porous lattice structures, but the smallest pore size to date has been about 50μm. This is due to the accuracy and precision of the laser as well as thermal stresses and heat spreading.

In this work a different approach was taken towards 3D printed wick structures. Instead of building a defined lattice structure, the laser power, speed, and spacing was varied to partially sinter the metal powder together instead of fully melting the particles. This results in a wick structure very similar to that of traditionally sintered wicks.

2. Experimental Methods

An experimental optimization study was completed by printing a total of 15 samples using a range of DMLS parameters. The goal was to achieve the smallest possible pore size to maximize capillary pumping power. For an LHP the capillary pressure in the primary wick must overcome the total pressure drop of the loop. Each sample was 2.54cm in length and 1.27cm in diameter. Each sample was printed using 316LSS on an EOSINT M280 machine. The parameters which were varied between samples includes laser power, speed, and spacing. By varying these parameters, the metal powder can be sintered together without forming a fully dense part leaving pores for fluid flow.

The pore size of each sample was measured using a bubble test. In the bubble test one side of the wick sample was submerged in methanol while the other side was attached to a pressurized nitrogen line. Methanol is used because it is compatible with stainless steel, and is safer to use for testing than ammonia. The pressure of the nitrogen was slowly increased until nitrogen began to bubble through the wick into the methanol. The pressure where bubbles began to form was recorded and the pore radius was calculated using Eq. 1.

$$r = \frac{2\sigma}{P} \qquad (1)$$

where r is the pore radius, σ is the surface tension of the fluid used, and P is the bubble point pressure.

A proof of concept LHP prototype was built as pictured in Figure 2 using a 3D printed primary wick. All wetted parts were 316LSS. The working fluid used was ammonia. The primary wick was 2.54cm in diameter and 10.2cm in length. The primary wick as printed had a porous interior with a fully dense envelope which was welded directly to the compensation chamber and vapor line. The tubing was 0.318cm in diameter and the condenser was 99cm in length. The sink temperature for testing was set to 0°C.

capillary pressure for an LHP in the 100-300W range. The DMLS parameters for sample 2 were used for fabrication of the LHP prototype. Achieving smaller pore sizes in likely only possible by using smaller diameter metal powder as the base material, but there are no current commercially available solutions.

Sample	Pore Radius (µm)
1	6.2
2	5.6
3	11.6
4	10.3
5	Hollow
6	Solid
7	Solid
8	Solid
9	6.0
10	8.8
11	Hollow
12	31.4
13	20.0
14	13.7
15	29.9

Table 1: Optimization study on DMLS parameters for creating primary wick

The steady state testing results from the LHP prototype are provided in Figure 3. Startup occurred almost instantly at a power of 110W which can be seen by a rapid decrease in the temperature of the liquid line. The power was increased in 5W increments, and steady state was achieved at a maximum power of 125W. At 130W the temperatures of the LHP began to increase linearly which is an indication of dry-out in the primary wick.

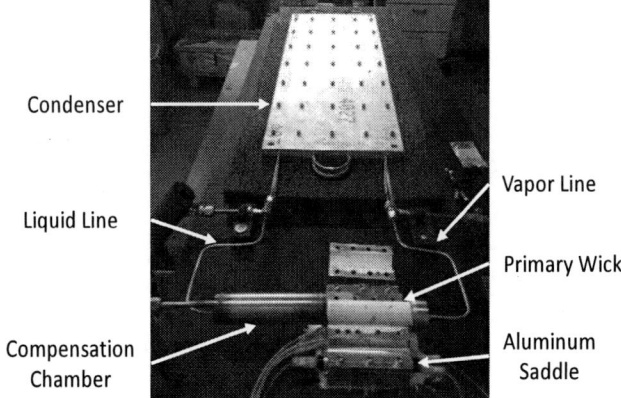

Figure 2: Completed LHP prototype with 3D printed primary wick.

3. Results

The results of the DMLS parameter optimization study are presented in Table 1. The smallest pore size achieved was 5.6µm from sample 2 which is capable of providing enough

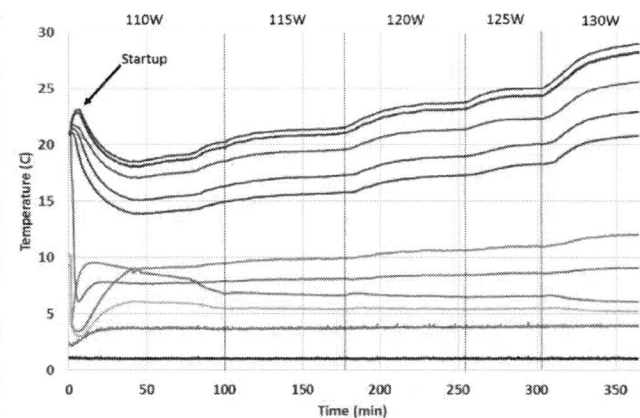

Figure 3: Steady state testing of LHP prototype

A temperature plot of the low power startup test is shown in Figure 4. A power of 5W was applied to the evaporator. The primary wick slowly heated up until there was sufficient superheating of the liquid ammonia in the vapor grooves to promote boiling. The ability to start the LHP at a power of only 5W indicates that the amount of heat leak is small, because significant heat leak would cause the primary wick and compensation chamber temperatures to increase together preventing LHP operation.

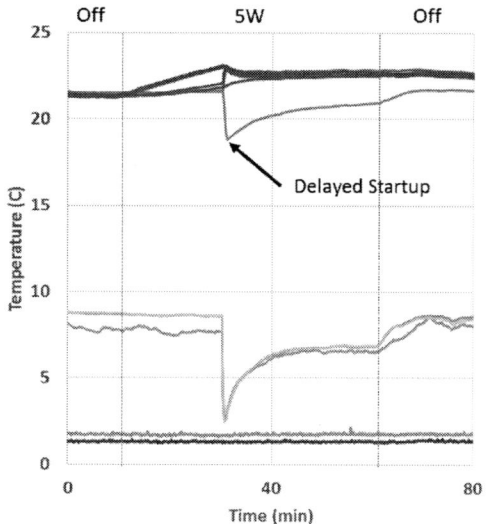

Figure 4: Successful startup of LHP prototype at a heat input of 5W as indicated by the rapid decrease in liquid line temperature.

Testing was completed with the evaporator raised 6in. above the condenser to verify the ability of the LHP to operate against gravity. A temperature plot of the results is presented in Figure 5. Startup and steady state operation were successful at powers of 50W and 70W.

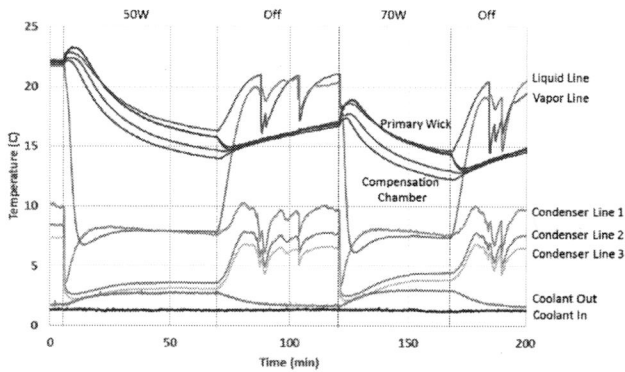

Figure 5: Test results with 6in. of adverse elevation between the evaporator and condenser. Steady state operation was reached at powers of 50W and 70W.

Power cycle testing was completed to verify the ability of the LHP prototype to handle transients. The power was rapidly changed between 70W and 20W. A temperature plot of the results is shown in Figure 6. Dry-out did not occur with rapid increases or decreases in power. This indicates that the secondary wick was able to maintain the supply of liquid to the primary wick during transient operation.

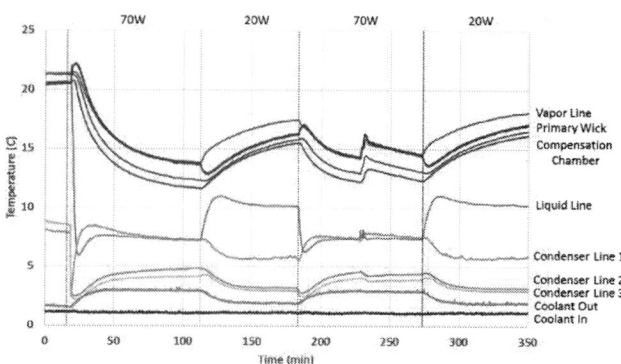

Figure 6: Rapid power cycling between 70W and 20W test results demonstrating the ability of the secondary wick to prevent primary wick dry-out during transients.

4. Conclusions

Significant progress has been made with the development of DMLS parameters for fabricating porous wick materials using 316LSS. An LHP prototype demonstrated the ability of a 3D printed wick to perform as an LHP primary wick. A maximum power of 125W was achieved which is in the power range of current CubeSat technologies. LHP performance was also verified for operation against gravity and during rapid changes in heat input power. Additional work is ongoing to optimize the 3D printed wick design as well as develop DMLS parameters for the secondary wick which has a larger pore size. This will eventually allow for the entire LHP evaporator to be 3D printed which will further reduce cost by eliminating more labor-intensive fabrication steps.

Acknowledgments

This work was funded by NASA through the Small Business Innovation Research (SBIR) program under contract NNX17CM09C. The technical monitor is Dr. Jeff Farmer.

References

1. Ku, Jentung. Operating characteristics of loop heat pipes. No. 1999-01-2007. SAE Technical Paper, 1999.

Research on Package Thermal Resistance of Power Semiconductor Devices

Koji Nishi

Ashikaga University

268-1, Omae-cho, Ashikaga-city, Tochigi, Japan

nishi.koji@v90.ashitech.ac.jp

Abstract

Power electronics is becoming more important than before with motor application expansion. For size reduction of inverter integrated motor design, accurate temperature prediction of power devices is becoming critical. For up to several hundred-watt motor system, inverter is designed with discrete power devices with standard package. This paper investigates package thermal resistance of a DPAK package as an example. Firstly, three-dimensional heat conduction simulation only with DPAK package model is conducted. It is found that its package thermal resistance changes by ~6.2°C/W due to boundary condition variation. After that, simulation not only with DPAK package but also with PCB is conducted to understand package thermal resistance of a real system implementation case. It is found that package thermal resistance is ~2.4 to 2.6 °C/W with smallest copper patterns, while its value is ~1.1 to 1.2 °C/W in the case that copper fully covers board as "layer".

Keywords

Power Electronics, Semiconductor, DPAK Package, Thermal Resistance, Thermal Management

1. Introduction

Power electronics is becoming more important than before, by demanding of power saving in consumer and industrial equipment. Also, system size reduction and integrated motor design is becoming common in many cases. For system size reduction, thermal design optimization is required and each thermal resistance along major heat transfer path(s) needs to be minimized. For this purpose, it is vital to understand and estimate each thermal resistance precisely. For system level investigation, surface temperature of major components and thermal solution is evaluated with both simulation and measurement. On the other hand, junction temperature of semiconductor devices is difficult to measure on real system and internal thermal resistance of power devices is not discussed so much. And many system designers estimate junction temperature from thermal resistance value on its product datasheet. However, heat transfer usually occurs with heat spreading and it causes thermal spreading resistance, which is known to change its value by boundary conditions. [1]

Thermal spreading resistance is defined as temperature difference per watt from smaller area to larger area along a face and is formulated as:

$$R_{th,spreading} = \frac{\bar{T}_{small} - \bar{T}_{large}}{\dot{Q}} \quad (1)$$

Here, $\bar{T}_{small}$ is average temperature of smaller area, $\bar{T}_{large}$ is average temperature of larger area and $\dot{Q}$ is heat transfer rate which passes though the face. Average temperature of an area can be defined as:

$$\bar{T} = \frac{1}{A} \int_A T dA \quad (2)$$

A simple example can be raised about aluminum alloy block of 50 mm × 50 mm × 5.0 mm, whose thermal conductivity is 209 W/mK (Figure1). Heat is applied uniformly along 10 mm × 10 mm area of the bottom center and whole top surface is cooled with boundary condition of the third kind which is governed with Newton's cooling law :

$$\dot{Q} = hA(T - T_{Ambient}) \quad (3)$$

Here, $\dot{Q}$ is heat transfer rate cooled down along the top surface, h is heat transfer coefficient, A is area of the top surface, T is temperature of the top surface, $T_{Ambient}$ is ambient temperature.

Thermal spreading resistance $R_{th,\ spreading}$ is defined along the bottom surface of aluminum alloy block, assuming heat input area as "smaller area" and whole bottom surface as "larger area", then, the value varies by heat transfer coefficient along the top surface (Figure 2). [2] While, thermal resistance of Aluminum alloy block is calculated from:

$$R_{th,material} = \frac{l}{kA} \quad (4)$$

Here, l is thickness, k is thermal conductivity and A is cross-sectional area. Thermal resistance of aluminum alloy block is 0.0096 °C/W, which is much smaller than thermal spreading resistance. This means thermal spreading resistance is dominant and highly affects on temperature rise of hot spot temperature.

The same thing can be said for power device package. That is, internal thermal resistance of power device can vary drastically with thermal spreading resistance if its boundary conditions, for example, PCB (Printed Circuit Board) size which mounts target power device, copper patterns of PCB, are changed. Usually, thermal resistance model is preferred, rather than three-dimensional simulation model for transient temperature prediction of power electronic systems such as inverter circuit because its carrier frequency is several to ~20 kHz and it will lead large number of simulation time step which requires large computational resources for three-dimensional simulation. However, thermal resistance variation by boundary conditions needs to be cared for thermal resistance model. Therefore, this research investigates thermal resistance variation and relationship between its value and boundary conditions by utilizing three-dimensional model and three-dimensional heat conduction simulation.

For up to several hundred-watt motor system, inverter is designed with discrete power devices with standard package. Especially, surface mount type package such as DPAK and D²PACK is widely used in the world. This paper explores internal thermal resistance variation by boundary condition, by utilizing DPAK package model as an example. Three-dimensional heat conduction simulation is conducted, then, result is discussed to understand well about internal thermal resistance variation of DPAK package.

2. Investigation of DPAK Package Thermal Resistance Variation

In this chapter, only DPAK package is modeled and simplified boundary conditions are employed to see the fundamental thermal behavior of DPAK package, assuming that heat only transfers from bottom surfaces of DPAK package. [3] The chtMultiRegionSimpleFoam solver in OpenFOAM v4.1 [4] is utilized as heat conduction simulation software. Its governing equation is three-dimensional heat conduction equation:

$$\frac{\partial}{\partial x}\left(k_x \frac{\partial T}{\partial x}\right) + \frac{\partial}{\partial y}\left(k_y \frac{\partial T}{\partial y}\right) + \frac{\partial}{\partial z}\left(k_z \frac{\partial T}{\partial z}\right) = 0$$

(5)

Here, T is temperature, k_x, k_y and k_z are thermal conductivity of x, y and z direction, respectively. Boundary conditions are set as adiabatic:

$$\frac{\partial T}{\partial x} = 0 \qquad \text{for } x \text{ direction} \qquad (6)$$

$$\frac{\partial T}{\partial y} = 0 \qquad \text{for } y \text{ direction} \qquad (7)$$

$$\frac{\partial T}{\partial z} = 0 \qquad \text{for } z \text{ direction} \qquad (8)$$

or boundary conditions of the 3rd kind:

$$k_z \frac{\partial T}{\partial z} = h\left(T - T_{ambient}\right) \qquad \text{for } z \text{ direction} \qquad (9)$$

Here, h is heat transfer coefficient of the boundary and $T_{ambient}$.

2.1. DPAK Package Model

DPAK package is one of standard surface mount packages and includes die, die attach, heat spreader, mold resin and several leads. In this paper, DPAK package with three leads is modeled to discuss internal thermal resistance of DPAK package (hereafter, package thermal resistance) (Figure 3). Wires from die to leads are not included in this model because they are too fine and have higher thermal resistance than other materials. heat spreader, mold resin and two leads (Lead1 and Lead3) of three leads are bottom surface of DPAK package, which is connected to PCB via soldering on usual implementation.

Dimensional summary and thermal conductivity of each part of DPAK package model is shown in Table 1. Its dimensions are determined by referring JEDEC TO-252F [5] and JEITA SC-63. [6] Also, thermal conductivity is

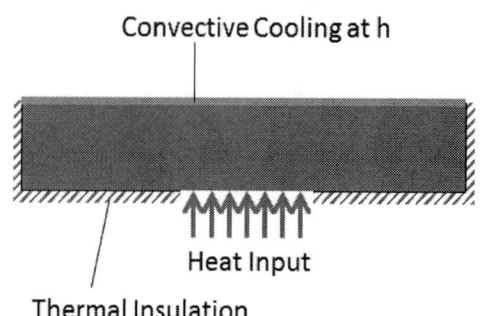

Figure 1: Side view of aluminum alloy block.

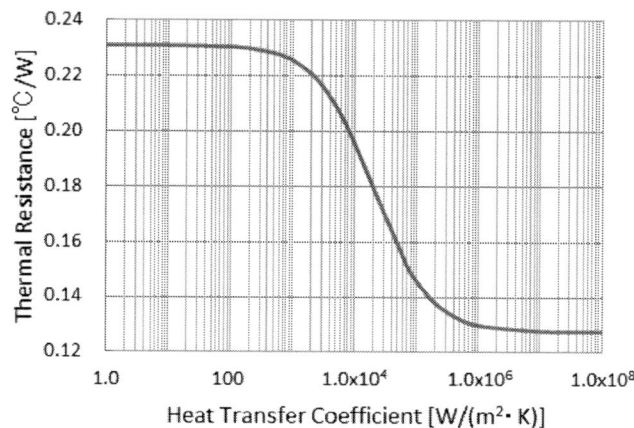

Figure 2: Thermal spreading resistance variation of aluminum alloy block.

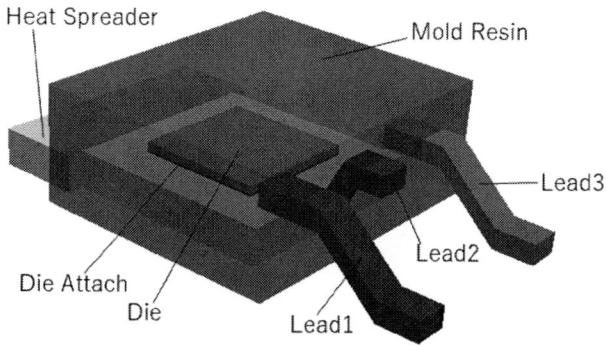

Figure 3: DPAK package model.

Part	Size [mm × mm × mm]	Thermal conductivity [W/mK]
Die	3.0 × 3.0 × 0.20	120
Die attach	3.0 × 3.0 × 0.050	50
Heat spreader	(5.8 × 2.0 + 5.0 × 5.0) × 0.90	300
Lead1/2/3	-	300
Mold resin	7.0 × 6.0 × 2.8	0.60

Table 1: Dimensions and thermal conductivity of DPAK package model.

determined by assuming die as silicon (Si), die attach as solder, heat spreader and leads as copper (Cu) alloy and mold resin as epoxy based material. This model is assumed to be vertical type Metal-Oxide-Semiconductor (MOS) transistor and whole die works as uniform heat source. Thus, uniform volume heat source is applied to whole die. Junction temperature ($T_{Junction}$) is defined as maximum temperature of the die.

2.2. Simulation Conditions

To understand how much package thermal resistance of DPAK package can vary by boundary conditions, only DPAK package itself in Figure 3 is modeled in this chapter. Thus, computational domain includes only DPAK package. Assuming that heat transfers via only bottom surface of DPAK package (i.e. bottom face of heat spreader, mold resin, lead1 and lead3), boundary conditions of the 3rd kind is applied to these bottom surfaces of DPAK package and all other surfaces are treated as adiabatic. Maximum grid size is set to 0.10 mm × 0.10 mm × 0.025 mm and total number of grids is ~0.9M.

2.3. Package Thermal Resistance of DPAK Package

Package thermal resistance of DPAK package is defined as:

$$R_{th,Junction-Bottom} = \frac{T_{Junction} - \overline{T}_{PkgBottom}}{\dot{Q}_{Bottom}} \quad (10)$$

Here, $\overline{T}_{PkgBottom}$ is average temperature of bottom surface in DPAK package and $\dot{Q}_{Bottom}$ is heat transfer rate from die to bottom surface of the DPAK package. $\dot{Q}_{Bottom}$ is the same as power dissipation of the die because it is assumed that heat only flows to bottom side of DPAK package for simplification in this paper.

2.4. Simulation Result

Package thermal resistance variation of DPAK package model is shown in Figure 4. At 1.0 W/m²K, package thermal resistance is ~6.7 °C/W. Increasing heat transfer coefficient value, package thermal resistance is decreased and reaches to ~0.5 °C/W at 1.0 × 10⁷ W/m²K. This means ~6.2 °C/W difference is caused by heat transfer coefficient difference. Usually, power semiconductor data sheets describe thermal resistance under specific condition. Also, if package thermal resistance is measured with almost isothermal surface environment such as large cold plate, heat transfer coefficient is theoretically ∞ and package thermal resistance will be near from 0.5 °C/W. Therefore, care must be taken if these values can represent for real system product design or not.

Temperature contours of DPAK package are shown in Figure 5. This shows thermal resistance value along center cross sectional face of DPAK package model from heat spreader to lead2, by dividing temperature rise by heat source value applied to die. Bottom surface temperature is different by each case because boundary conditions are different. Thus, these contour shows not absolute temperature value but temperature difference from junction temperature. Blue

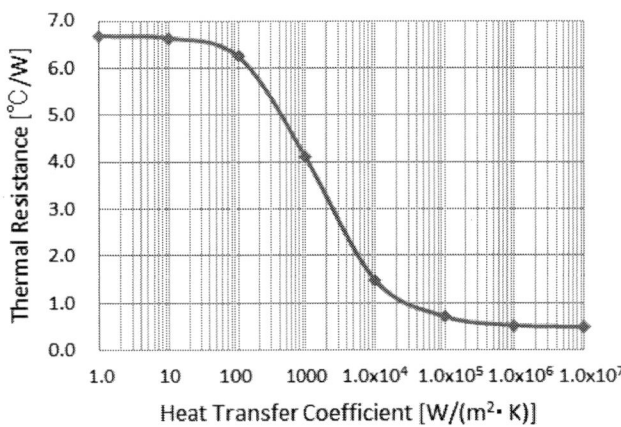

Figure 4: Package thermal resistance variation by heat transfer coefficient.

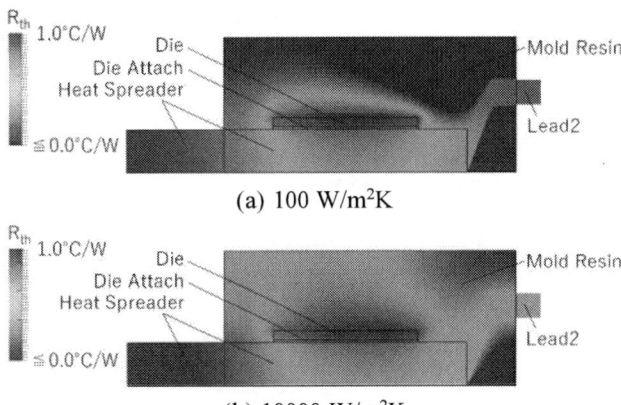

(a) 100 W/m²K

(b) 10000 W/m²K

Figure 5: Temperature contour by 1W of DPAK package model.

colored area of 100 W/m²K case is lager than that of 10000 W/m²K case in top right region. This is because much more heat passes through top right region for 100 W/m²K case, which causes larger temperature difference. In contrast, blue colored area of 10000 W/m²K case is lager than that of 100 W/m²K case in bottom left (heat spreader) region. This is because higher heat transfer coefficient along bottom surface of DPAK package allows heat to go smoothly through just beneath silicon die.

3. Investigation for Simplified DPAK Package with PCB

In this section, DPAK package model is evaluated with PCB (Print Circuit Board) to understand package thermal resistance variation by PCB implementation conditions. Also, simulation trials are conducted to understand its thermal resistance variation by simulation model construction difference. In this chapter, 6sigmaET [7] 12 is utilized as heat conduction simulation software. Its governing equation is 3-dimensional heat conduction equation and equations of boundary conditions are the same as that of chapter 2 (eq. (5) to (9)).

3.1. Simulation Model and Boundary Conditions

Figure 6 depicts model region in this chapter, which consists of DPAK package and PCB. DPAK package model is identical to one employed in chapter 2. PCB consists of board, copper patterns and soldering. copper patterns and soldering are located at only top surface of PCB. Board horizontal size for each case is 20 mm × 20 mm, 30 mm × 30 mm, 40 mm × 40 mm or 50 mm × 50 mm. DPAK package is placed so that its die center is located at center of PCB. Horizontal dimensions of copper pattern and soldering are different by each section below and are explained later. Thickness and thermal conductivity of each PCB component is shown in Table 2. Base material of PCB (board) is assumed as FR4. As for boundary conditions, boundary condition of the 3rd kind is applied to the bottom surface of PCB and all other surfaces are treated as adiabatic condition. Heat transfer coefficient of PCB bottom surface is 10 W/m²K. Maximum grid size is set to 0.10 mm × 0.10 mm × 0.025 mm. Uniform volume heat source is applied to whole die. Heat source is set to 1.0W to refer to temperature rise as thermal resistance in this chapter. Junction temperature ($T_{Junction}$) is defined as maximum temperature of the die.

3.2. Simulation with The Smallest Copper Pattern and Soldering

To understand package thermal resistance with typical implementation of DPAK package into PCB, copper pattern and soldering are constructed as shown in Figure 7 and Table 3. To know the worst case scenario of package thermal resistance, copper pattern area is set to almost the smallest size. Also, soldering area is the same as bottom surface area of heat spreader, lead1 and lead3.

Heat conduction simulation results is shown in Figure 8. Package thermal resistance is ~2.4 to 2.6 °C/W, depending on PCB size. Increasing PCB size, package thermal resistance of DPAK package slightly increases its value.

3.3. Simulation Result with Fully Covered Copper Layer

To understand the best case of package thermal resistance with DPAK package and PCB, this section investigates DPAK package and PCB with the maximum sized copper pattern. Different from previous section, copper pattern fully covers board surface as "layer". All other conditions are the same as previous section.

Heat conduction simulation result is shown in Figure 9. Package thermal resistance is ~1.1 to 1.2 °C/W, depending on PCB size. Increasing PCB size, package thermal resistance of DPAK package slightly increases its value. Compared to results in previous section, package thermal resistance is almost half and copper pattern highly affects on package thermal resistance.

3.4. Simulation Result without Soldering

To understand the influence of model components to package thermal resistance, this section investigates DPAK package and PCB without soldering area. DPAK package

Figure 6: Whole view of simulation model.

Part	Thickness [mm]	Thermal conductivity [W/mK]
Board	1.6	0.40
Copper patterns / layer	0.035	400
Soldering	0.10	50

Table 2: Thickness and thermal conductivity of PCB components.

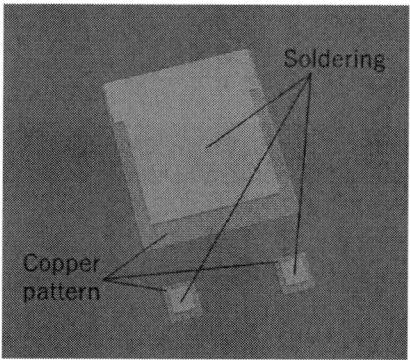

Figure 7: Copper pattern and soldering area of simulation case for typical implementation.

Part	Copper Pattern [mm×mm×mm]	Soldering Area [mm×mm×mm]
Heat spreader	6.0 × 8.0 × 0.035	(5.8 × 2.0 + 5.0 × 5.0) × 0.10
Lead1/3	3.0 × 3.0 × 0.035	0.8 × 1.0 × 0.035

Table 3: Dimensions of copper pattern and soldering area.

model is directly connected to copper patterns. All other conditions are the same as section 3.2 including copper pattern size (the smallest size copper patterns).

Heat conduction simulation result with the smallest copper patterns and without soldering is shown in Figure 10. Package thermal resistance is 2.0 to 2.1 °C/W, depending on PCB size. Increasing PCB size, package thermal resistance of DPAK package slightly increases its value. Compared to results in section 3.2, package thermal resistance is lower by 0.4 to 0.5 °C/W. Therefore, soldering area modeling has some effects on temperature prediction.

4. Conclusions

This paper investigates package thermal resistance variation of DPAK package by boundary condition, utilizing simplified model of only DPAK package and DPAK package with PCB. Major findings are as follows:

- Package thermal resistance of DPAK package varies drastically by bottom boundary condition. Its value is changed from ~0.5 to ~6.7 °C/W in the case that boundary conditions of the 3rd kind is applied to its bottom surface.
- In the case with the smallest copper pattern and soldering, package thermal resistance of DPAK package is ~2.4 to 2.6 °C/W. While, in the case with fully covered copper layer and soldering, it is ~1.1 to 1.2 °C/W. This means copper pattern affects a lot for package thermal resistance of DPAK package.
- Increasing PCB size, package thermal resistance of DPAK package slightly increases its value.
- If soldering component is removed in simulation model, package thermal resistance becomes lower by 0.4 to 0.5 °C/W.

In this paper, it is assumed that heat only transfers via only bottom surface of DPAK package. Further investigation is required for the case that heat transfers to not only bottom surface but also other surfaces as the next step.

References

1. Muzychka, Y. S., Culham, J. R. and Yovanovich, M. M., "Thermal Spreading Resistance of Eccentric Heat Sources on Rectangular Flux Channels", Journal of Electronic Packaging, Vol. 125, pp. 178–185, 2003.
2. Nishi, K. Imano, M., "Validation of OpenFOAM Steady-State Solver with Conjugate Heat Transfer between Solid and Fluid Regions, targeting for Heat Conduction Problem", Journal of The OpenCAE Society of Japan, Vol.1, No.1, OpenCAE2018-001, 2018 (In Japanese).
3. Nishi, K., "Research on package thermal resistance of power semiconductor devices Investigation of DPAK package", 55th National Heat Transfer Symposium of Japan, K231, 2018 (In Japanese).
4. OpenFOAM v4.1, https://openfoam.org/version/4-1/ (Access Date: 2018/10/12).
5. "FLANGE MOUNTED FAMILY SURFACE MOUNT (PERIPHERAL TERMINALS)", JEDEC, TO-252F, 2017.
6. "Standard Outlines of Semiconductor Devices (Discrete Semiconductor Devices)", JEITA ED-7500B, pp. 19-20, 2015.
7. 6sigmaET, https://www.6sigmaet.info (Access Date: 2018/10/12).

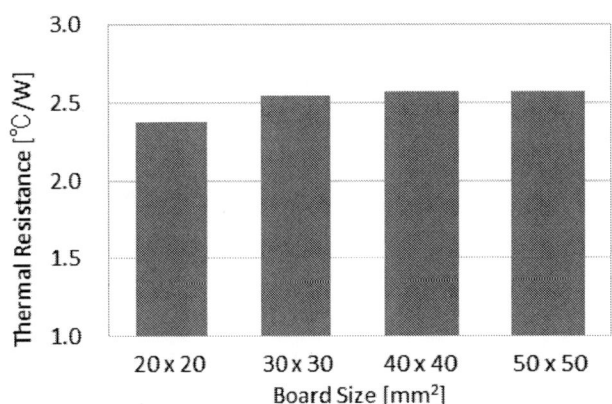

Figure 8: Thermal resistance variation by board size with the smallest copper pattern and soldering.

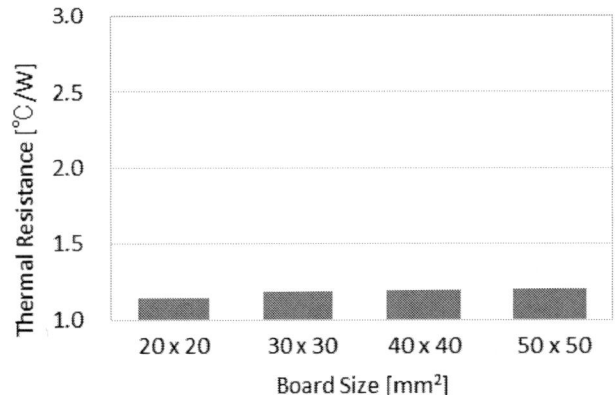

Figure 9: Thermal resistance variation by board size with full copper layer and soldering.

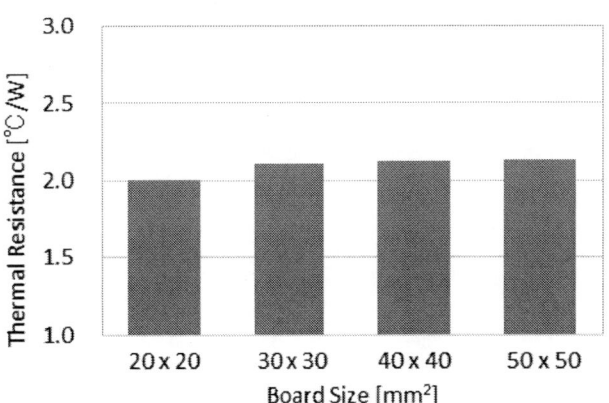

Figure 10: Thermal resistance variation by board size with the smallest copper pattern only.

An Ultra-Thin Loop Heat Pipe with Long-Distance Heat Transport for Cooling of Small Electronic Devices

Shuto Tomita, Ai Ueno, Hosei Nagano
Nagoya University
Furo, Chikusa, Nagoya, Aichi, Japan
Tomita.shuto@h.mbox.nagoya-u.ac.jp

Abstract

In this paper, an ultra-thin loop heat pipe is proposed for small electronic devices. The thin LHP (TLHP) was designed, fabricated, and tested to meet the heat dissipation requirements of small or thin electronic devices. The proposed LHP with an evaporator thickness of 1.0 mm has a one-way heat transport length of 200mm, and a heat transport capability above 10 W. The test results showed that the TLHP could be transferred up to 12 W and maximum temperature of the heater is about 90 °C. The TLHP showed stable operation at different power cycles with a heat load of 5-10 W. Hysteresis is not confirmed after three cycles of the heat load.

Keywords

Loop heat pipes, electronics cooling, two-phase cooling

1. Introduction

In recent years, thermal management is essential for thin or small electronic devices such as smartphones, tablets, and laptops. Furthermore, thermal design power (TDP) of CPU or GPU is rising year by year. So far, high thermal conductivity materials such as graphite sheets were used for heat dissipation. Moreover, a heat pipe with the thickness of under 0.5 mm is used for heat dissipation in smartphones. For future applications, it is expected to pursue the possibility of high-performance heat transport technology as the next generation of thermal management.

Loop heat pipe (LHP) is a two-phase heat transfer device that utilizes the evaporation and condensation of the working fluid to transfer heat; the LHP uses capillary force in porous wick to circulate the fluid. Figure 1 shows the schematic of the LHP. The LHP has been used for thermal control of spacecraft and satellites [1, 2].

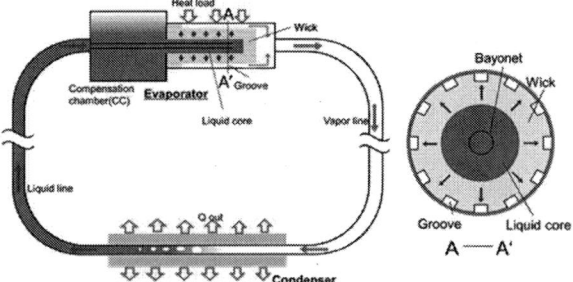

Figure 1: Schematic of the LHP.

In recent years, research and development of the LHP for terrestrial applications have been conducted. Under such situation, several studies on small LHPs exist. Chen et al. fabricated an LHP with a 5 mm thick cylindrical evaporator [3].

The shapes of conventional LHPs are cylindrical, and their thermal resistances between the evaporator and heat source become large due to the incomplete contact area. The structure of a flat type evaporator is suitable for reducing the thermal resistance between the evaporator and heat source. However, further thinning of the evaporator is required for mounting applications in small electronic devices such as tablets and laptops. Zhou et al. fabricated a 1.2 mm thick LHP with a flat type evaporator. The LHP has a one-way transport length of 105 mm and a heat transport amount of 12 W [4]. Shioga et al. fabricated an LHP with a 0.6 mm thick evaporator, and the results showed that the LHP could be transferred up to 15 W [5]. Although several studies on small LHPs exists as mentioned above, just a few studies have focused on thin LHPs (TLHPs) with long heat transport length.

In our previous study, a micro LHP (mLHP) with a new evaporator structure was proposed and fabricated. The thickness of the evaporator in the mLHP was 3 mm, and the heat transport capability was 11 W [6]. In this paper, an improved LHP with the evaporator thickness of 1.0 mm is proposed. The TLHP has a one-way transport length of 200 mm, the evaporator size of 30 mm × 30 mm with a thickness of 1.0 mm and a heat transport capability above 10 W. SHIRASU Porous Glass (SPG) which has low thermal conductivity and high porosity are used as a wick in the evaporator to reduce heat leak.

2. Evaporator structure

Figure 2 shows the designed evaporator case made by a 3D-printer which informs grooves, liquid cores, and ribs. For increasing the transport distance, the large volume of CC is needed, because the size of the CC depends on the size of vapor and condenser lines. Ribs are installed in the evaporator to support the structure because the thinnest part of the evaporator with a thickness of 0.20 mm is significantly affected by thermal distortion when fabricated with a 3D printer.

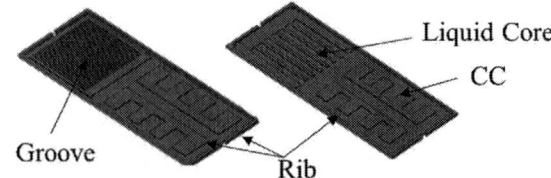

Figure 2: Schematic view of the evaporator case (left: a top side, right: a bottom side)

Figure 3 shows the sectional view of the evaporator. Wick is in the form of a plate and does not have special processing; it is enclosed inside of the evaporator.

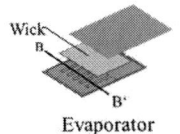 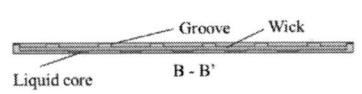

Figure 3: Schematic view and sectional view of the evaporator

Figure 4 shows how the fluid path when passing through the wick is shortened by installing the liquid core. In our previous study, liquid cores to assist the supply of working fluid to the wick. Furthermore, by installing the core under the wick, the path length of the working fluid in the wick can be shortened. Therefore, the pressure drop of the working fluid can be reduced [6]. The ratio of heat transfer area and groove area is 1:1, and designed structure promote evaporation because of increasing the number of liquid-vapor interface [7]. Shirasu Porous Glass (SPG) was used as the wick. They can reduce the heat leak through the wick because they have low thermal conductivity. Table 1 shows properties of wick materials which were measured in our laboratory. Figure 5 shows SEM images of SPG wick. Table 2 shows geometric characteristics of LHP. The temperature distribution of TLHP during the experiment was measured with 12 T-type thermocouples. The power was applied with a ceramic heater with the size of 20 mm × 20 mm. Figure 5 shows a photo of the TLHP. Ethanol was selected as the working fluid because its operating temperature is low, and it is non-toxic. The vapor line, liquid line, and condenser are all made using 1/16-inch (1.59 mm) tubes with the length of 200 mm. The condenser is formed using an Al5052 aluminum plate with the size of 100 mm × 200 mm × t 3mm.

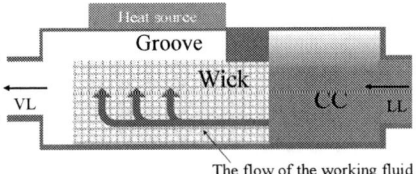

The flow of the working fluid

(a)without liquid core

The flow of the working fluid

(b)with liquid core

Figure 4: The flow of the working fluid in the evaporator(a)without liquid core (b)with liquid core

Table 1: Specifications of SPG

Mean pore radius	4.0 μm
Porosity	60 %
Permeability	$2.0 \times 10^{-13} \, m^2$
Thermal Conductivity	1.38 W/(mK)
Heat resist temperature	550 °C

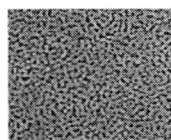

Figure 5: SEM images of the wick made from SPG

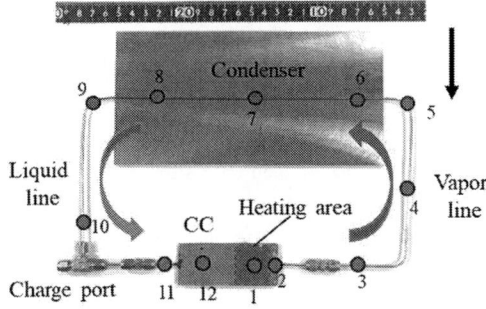

Figure 6: Photograph of TLHP

Table 2: Geometric characteristics of TLHP

Evaporator size	28 × 30 × t1.0 mm
Evaporator case thickness	0.2 mm
Wick	25 × 25 × t0.30 mm
CC	28 × 43 × t1.0 mm
Vapor line (I.D.)	0.98 mm
Vapor line (O.D.)	1.59 mm
Vapor line (length)	200 mm
Condenser line (I.D.)	0.98 mm
Condenser (O.D.)	1.59 mm
Condenser line (length)	200 mm
Condenser Size	100 × 200× t3.0 mm
Liquid line (I.D.)	0.98 mm
Liquid line (O.D.)	1.59 mm
Liquid line (length)	200 mm
Heating area	20 × 20 mm
Working fluid	Ethanol
Bubble point	7 kPa

3. Results

The experimental conditions under which the TLHP performance is evaluated are listed in Table 3.

Table 3: Experimental conditions

	Power step-up	Power cycling
Heat load history	2w – 12 W	5W / 10W
Time interval	600 sec.	900 sec.
Cooling condition	Air natural cooling	
LHP attitude	Bottom heat	

The power step-up experimental results representing the temperature profile are shown in Figure 7. T_{ec} is the temperature of the evaporator (No.2 in Figure 6), T_{cc} is the temperature of the CC (No.12 in Figure 6) and $T_{con,inlet}$ is the temperature of the condenser inlet (No.6 in Figure 6). The CC temperature does not change substantially until the heat load at 5 W, and it starts rising as the heat load moves upward from 6 W. The heater temperature is under 100 °C at 14 W.

Figure 8 shows the results of the evaporator and the CC temperatures, and thermal resistance in a steady state. The thermal resistance is calculated as,

$$R_{LHP} = \frac{T_{ec} - T_{con}}{Q_{load}} \tag{1}$$

Here, T_{ec} is the temperature of the evaporator, T_{con} is the average temperature of the condenser (No.6, 7, 8 in Figure 6), and Q_{load} is the heat load. The TLHP developed in this study is capable of transporting heat loads of up to 12 W, and temperature reaches the heatproof temperature at 14 W. T_{ec} is 83.0 °C at 12 W, and the thermal resistance is the smallest at 1.98 °C/W.

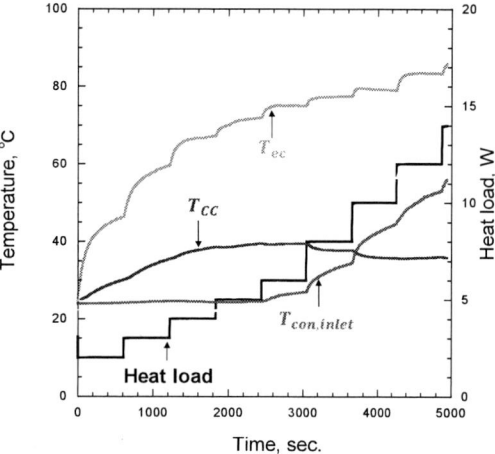

Figure 7: Temperature profile of the power step-up test

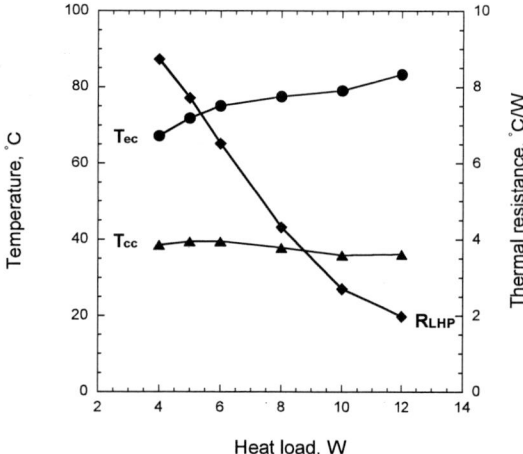

Figure 8: Effects of heat load on the evaporator temperature, the CC temperature and the thermal resistance.

The TLHP is tested at different power cycles with a heat load of 5-10 W and an interval of 900 sec.; three cycles are used to confirm its work stability against heat generation changes. The experimental results are shown in Figure 9. T_{heate} is the

temperature of the heater (No.1 in figure 6). The operating temperature (T_{ec}) is approximately 72 - 78 °C. Hysteresis is not confirmed after three cycles of the heat load, and stable operation is confirmed for the target heat load.

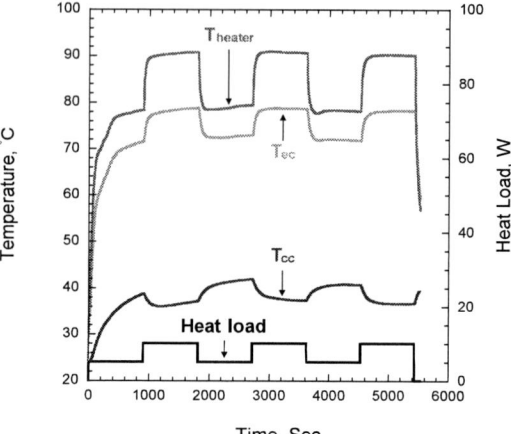

Figure 9: Effects of power cycle on the heater, evaporator and CC temperature of the TLHP

4. Conclusions

In this study, with the aim to adapt an LHP to thin electronic devices such as tablets and laptops. The ultra-thin LHP is designed and fabricated, and experiments are performed. The conclusions are summarized as below:

- A TLHP with a thickness of 1 mm and the heat transport length of 200 mm is designed and fabricated.
- The TLHP can be transferred up to 12 W, and the temperature is 83 °C and thermal resistance is 1.98 °C/W.
- The TLHP showed stable operation at different power cycles with a heat load of 5-10 W. Hysteresis is not confirmed after three cycles of the heat load.

Acknowledgments

Place acknowledgments here, if needed.

References

1. Maydanik, Yu F., Loop heat pipes, Applied Thermal Engineering, Volume: 25, Issue: 5, pp. 635-657, 2005
2. Ku, J., Operating characteristics of loop heat pipes, SAE Technical Paper, No. 1999-01-2007, 1999
3. Maydanik, Yu.; Vershinin, S.; Chernysheva, M.; Yushakova, S., "Investigation of a compact copper–water loop heap pipe with a flat evaporator.", Applied Thermal Engineering, Volume: 31, Issue: 16, pp. 3533-3541, 2011
4. Zhou, G.; Li, J.; Lv, L.; Peterson, G. P., "Comparative Study on Thermal Performance of Ultrathin Miniature Loop Heat Pipes With Different Internal Wicks.", Journal of Heat Transfer Volume: 139, Issue: 12, 2017
5. Shioga, T.; Mizuno, Y., "Micro Loop Heat Pipe for Mobile Electronics Applications.", SEMI-THERM 2013, 2013

6. Fukushima, K., Nagano, H., New evaporator structure for micro loop heat pipes. International Journal of Heat and Mass Transfer, Volume: 106, pp. 1327-1334, 2017

7. Odagiri, K.; Masahito N.; Hosei N., "Microscale infrared observation of liquid-vapor interface behavior on the surface of porous media for loop heat pipes." Applied Thermal Engineering, Volume: 126, pp. 1083-1090, 2017

Relative Performance of Two-Phase vs Solid Conductive Heat Spreaders for High Heat Flux Applications

Mr. Joe Boswell, Dr. Corey Wilson, Mr. Josh Schorp, and Mr. Dan Pounds
ThermAvant Technologies, 1000A Pannell Street, Columbia, MO 65201
(573) 397-6912, joe.boswell@thermavant.com

Dr. Bruce Drolen
Engineering Consultant to ThermAvant Technologies

SUMMARY

State of the art integrated circuit devices operate at ultra-high heat fluxes with local hotspots generating well above 1,000 W/cm². Heat spreaders attach to devices and transform their heat fluxes to lower levels for rejection to heat sinks with given surface areas, heat transfer coefficients and boundary temperatures. Thermal engineers design heat spreaders to minimize device hotspots by optimizing material selection, geometry, and heat spreader type. Heat spreader types include solid conductive units as well as two-phase solutions; i.e., wick-based heat pipes (or vapor chambers) and pressure-driven oscillating heat pipes (OHPs). Two-phase spreaders more evenly diffuse heat across the heat sink but also add superheat at the interface of their solid wall and working fluid (i.e., convection boundary). For two-phase spreaders, thinner walls lower the temperature rises through the wall but increase heat fluxes, superheats and thus temperature rises at the convection boundary.

The goal of the following study is to establish the optimum wall thickness of a heat spreader that decreases the temperature rise from the device through the spreader to the convection boundary – or in the case of a solid conductor to the heat sink boundary. In this paper, exact conduction solutions for this optimization problem are presented for both rectangular and radial geometries. For corroboration, the results are compared to finite element solutions for several sample problems with excellent agreement. The utility of the solutions is that they can readily be used in a spreadsheet format for rapid thermal trades to identify the optimum heated wall thickness and provide the minimum device temperature.

1. INTRODUCTION

The primary problem addressed in this paper is that of minimizing the temperature of a high heat flux source on a heat spreader with an internal two-phase working fluid, e.g., wick-based vapor chamber or pressure-driven oscillating heat pipe. The goal is to determine the optimum wall thickness of such a heat spreader to decrease the temperature rise through the wall without excessive temperature build-up at the wall's interface with the internal working fluid.

Guo, et al. [1] address the problem of a heat source of flux q_o with radius a on a heat spreader with thermal conductivity k, radius L and thickness H. They use a constant temperature boundary condition for their solution. They derived an exact solution using the separation of variables method and compared this solution to a much simpler approximate solution they termed a composite solution. Although their solution provides guidance, the isothermal boundary condition causes the temperature rise (ΔT) to simply increase with increased wall thickness.

Lee [2] presents a correlation of numerical results for spreading of heat from a rectangular source and heat loss by convection from the opposite face that is useful for problems such as those addressed here, but it is not an exact solution, and it is difficult to base an optimization off of such a correlation. Gholami and Bahrami [3] present the solution for a plate with an arbitrary number of prescribed rectangular heat flux inputs and outputs. They do not include the case of a convective thermal boundary. The solutions that are presented are quite general but leave significant work to yield the final solution for any given case. Thompson and Ma [4] present results for heat spreading in a rectangular geometry with convective heat loss from the opposite side of the plate. Unfortunately, the result is only accurate for Biot numbers less than 0.001. The intent was to simulate spreading in two-phase spreaders when modeled as an effective conductor with very high thermal conductivity; thus, the small Biot number limitation was appropriate in this case. For the model presented here, the h-value is high due to two-phase heat transport within the spreader and the wall conductivity will be far less than the effective thermal conductivity of an OHP, vapor chamber or heat pipe. Thus, a solution is needed that does not require the limit of small Biot number for addressing the trade of thickening the heated wall of a two-phase spreader to minimize the resultant component temperature.

2. ANALYTICAL SOLUTION – RECTANGULAR GEOMETRY

In many high heat flux applications, the heat source is one or more electrical components attached to a heat spreader configured in a linear configuration as shown in Figure 1. In the case of a two-phase heat spreader, the convection boundary condition is assumed to represent the heat transfer from the heat spreader's wall into its internal two-phase working fluid.

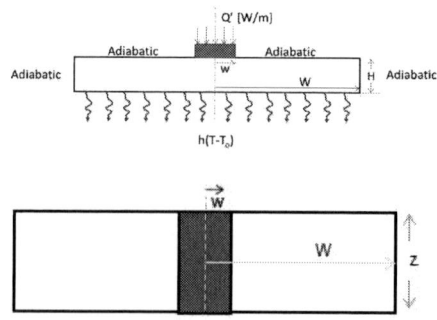

Figure 1 – Configuration of high heat flux device on OHP surface (left, side view; right top view)

The analysis that follows is rectangular and is considered infinite in the direction aligned with the heat source, Figure

2. In the best case where w = W or where the thermal conductivity of the heat spreader is infinite, the thermal resistance is just the sum of the conduction resistance through the spreader and the convection resistance to the sink. This limiting thermal resistance is shown below

$$R_{cond} = \frac{H}{kW} \left[\frac{mC}{W}\right] \quad and \quad R_{conv} = \frac{1}{hW} \left[\frac{mC}{W}\right]$$

The temperature rise that results from this net resistance and a heat flow of Q' is

$$\Delta T = \frac{Q'}{2W} \left(\frac{H}{k} + \frac{1}{h}\right)$$

In the worst case, where the heat spreader is infinitely thin (or non-existent) the maximum temperature rise is

$$\Delta T = \frac{Q'}{2wh}$$

The exact solution derived to this problem in the analysis that follows agrees with these two limiting cases.

The solution will be derived using the method of separation of variables. Figure 2 shows the reduced geometry, taking advantage of the symmetry condition at x=0.

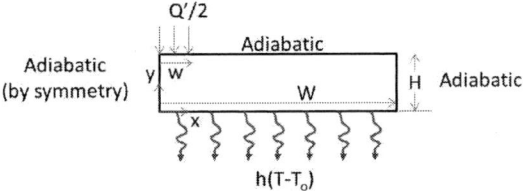

Figure 2 – Coordinate system for separation of variables solution

By introducing θ= (T-T$_o$) the differential equation to be solved is

$$\frac{\partial^2 \theta}{\partial x^2} + \frac{\partial^2 \theta}{\partial y^2} = 0$$

$$\frac{\partial \theta(0,y)}{\partial x} = 0, \quad \frac{\partial \theta(W,y)}{\partial x} = 0, \quad \frac{\partial \theta(x,0)}{\partial y} = \frac{h}{k}\theta,$$

$$and \quad \frac{\partial \theta(x,H)}{\partial y} = \frac{Q'}{2wk} f(x)$$

$$where \quad f(x) = 1, when \; 0 \le x \le w, and \; f(x)$$
$$= 0, when \; w < x \le W$$

Following the method of separation of variables, assume a product solution of the form

$$\theta(x,y) = X(x)Y(y)$$

Substitution yields

$$\frac{1}{X}\frac{d^2 X}{dx^2} = -\frac{1}{Y}\frac{d^2 Y}{dy^2} = \pm\lambda^2$$

Since the x-axis is the homogeneous direction the sign of λ^2 is chosen to be negative to yield a characteristic value problem. The general solution of the characteristic value problem in the x-direction is

$$X(x) = C_1 sin\lambda x + C_2 cos\lambda x$$

Introducing the boundary conditions in the x-direction gives

$$\lambda_n = \frac{n\pi}{W}, \; n = 0,1,2,3,\dots \; and \; X(x) = C_2 cos\frac{n\pi x}{W}, \; n = 0,1,2,3,\dots$$

The solution in the y-direction is given in terms of the hyperbolic functions cosh and sinh

$$Y(y) = C_3 \left(cosh\lambda y + \frac{h}{k\lambda} sinh\lambda y\right)$$

Combining the solutions for X(x) and Y(y) yields

$$\theta(x,y) = C_0 \left(1 + \frac{hy}{k}\right) + \sum_{n=1}^{\infty} \left[C_n \left(cosh\frac{n\pi y}{W} + \frac{hW}{kn\pi} sinh\frac{n\pi y}{W}\right) cos\frac{n\pi x}{W}\right]$$

The nonhomogeneous heat input boundary condition is introduced to derive the coefficients, C_n.

$$C_0 = \frac{Q'}{2hW}, \quad and \quad C_n = \frac{Q'}{wk} \frac{\frac{1}{n\pi} sin\left(n\pi\frac{w}{W}\right)}{\left(\frac{n\pi}{W} sinh\frac{n\pi H}{W} + \frac{h}{k} cosh\frac{n\pi H}{W}\right)}$$

Substitution into the expression for θ(x,y) results in the final solution

$$\theta(x,y)$$
$$= \frac{Q'}{2W} \left\{ \frac{1}{h} + \frac{y}{k} \right.$$
$$+ \frac{W}{wk} \sum_{n=1}^{\infty} \left[\frac{\left(cosh\frac{n\pi y}{W} + \frac{hW}{kn\pi} sinh\frac{n\pi y}{W}\right)}{\left(\frac{n\pi}{W} sinh\frac{n\pi H}{W} + \frac{h}{k} cosh\frac{n\pi H}{W}\right)} \left[\frac{2}{n\pi} sin\left(n\pi\frac{w}{W}\right)\right] cos\frac{n\pi x}{W} \right] \left. \right\}$$

The temperature at the hottest location directly under the heat source, i.e. the "hotspot", is given as follows

$$T(0,H)$$
$$= T_{\infty}$$
$$+ \frac{Q'}{2W} \left\{ \frac{1}{h} + \frac{H}{k} + \frac{W}{k} \sum_{n=1}^{\infty} \left[\left(\frac{1 + \frac{hW}{kn\pi} tanh\frac{n\pi H}{W}}{\frac{hw}{k} + n\pi\frac{w}{W} tanh\frac{n\pi H}{W}}\right) \left[\frac{2}{n\pi} sin\left(n\pi\frac{w}{W}\right)\right] \right] \right\}$$

The heat flux at the heat rejection surface, y = 0, is given below.

$$q_o = k\frac{\partial\theta(x,0)}{\partial y} = \frac{Q'}{2} \left\{ \frac{1}{W} + \frac{1}{w} \sum_{n=1}^{\infty} \left[\frac{\frac{2}{n\pi} sin\left(n\pi\frac{w}{W}\right) cos\frac{n\pi x}{W}}{\frac{kn\pi}{hW} sinh\frac{n\pi H}{W} + cosh\frac{n\pi H}{W}} \right] \right\}$$

The goal of this study was to determine the optimum thickness, H, to provide the best heat spreading with the least added temperature rise through the thickness. However, a closed form solution of the resultant equation for the optimum thickness is not possible due to the transcendental functions in the solution. The optimum can easily be determined using a spreadsheet formulation of the above hotspot temperature at x=0 and y=H.

3. RESULTS – RECTANGULAR GEOMETRY

Table 1 shows a set of parameters for a notional application with 60 W heat-generating device sized at 2.0 mm x 1.6 mm (1,909 W/cm² input heat flux) centrally located on a 12.5 mm x 1.6 mm two-phase heat spreader (12:1 surface area ratio of spreader to device). The h-value at the two-phase convection boundary is input as 50,000 W/m²K. The chosen material was Copper (Cu) for its relatively high thermal conductivity, despite its coefficient of expansion being greater than those of typical electronics. The initial wall thickness was set at 1 mm.

71

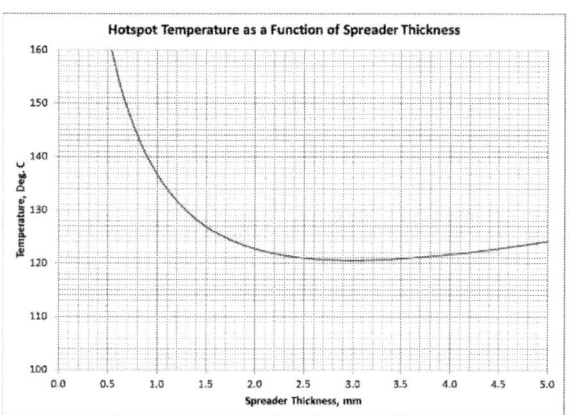

Total Power of Heat Source =	60	W		
Length of Heat Source (z) =	0.001571503	m	1.6	mm
Q' (total to the part) =	38180.0	W/m		
Heat Flux at Source =	**1909.0**	**W/cm^2**		
H =	0.001	m	1.0	mm
W (half of part) =	0.0125	m	12.5	mm
w (half of heat input width) =	0.001	m	1.0	mm
k =	400	W/m-K		
h =	50000	W/m^2-K		
hH/k =	0.1250	[ ]	Biot number	
$T_{inf}=T_{v,OHP}$ =	0	C		
$T_{inf}+Q'/2W*(H/k+1/h)$ =	34.4	C		
$T_{inf}+Q'/2Wh$ =	30.5	C	Minimum Temperature	
$T_{inf}+Q'/2wh$ =	381.8	C	Maximum Temperature	
$T_{hotspot}$ =	136.6	C	**Hotspot Temperature**	

Table 1 – Geometric and other input parameters for a notional high flux electronics heat spreader application

Figure 3 shows the predicted hotspot temperature at the interface between the device and the two-phase heat spreader to be 137°C above the working fluid's boundary temperature for a thickness of 1 mm. The optimum thickness is predicted to be about 3.0 mm for the wall of the vapor chamber or OHP beneath the electrical components which lowers the maximum temperature to 120°C. Note that reducing the thickness below approximately 1 mm has a severe impact on the peak temperature while increasing thicknesses above 3.0 mm is incrementally less harmful.

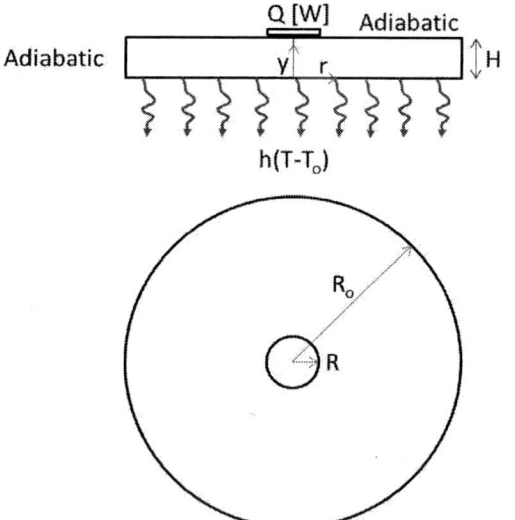

Figure 3 – Predicted hotspot temperatures of a notional device vs. the thickness of the two-phase spreader wall.

Figure 4 shows the predicted heat flux out from the heat spreader's wall (i.e., cooled surface or fluid boundary) at the optimal wall thickness – which in this case is 3mm. The peak heat flux has been transformed 1909 W/cm² to just over 400 W/cm² as the heat traverses through the walls thickness.

Using this same analytical method, if the same 1,909 W/cm² device were attached to a solid Cu heat spreader that rejected heat to a single-phase fluid (e.g., boundary h-value of 5,000 W/m²-K) in place of the two-phase Cu heat spreader, then the hot spot would be optimized to 402°C above the fluid boundary at a spreader thickness (H) of about 4.5 mm.

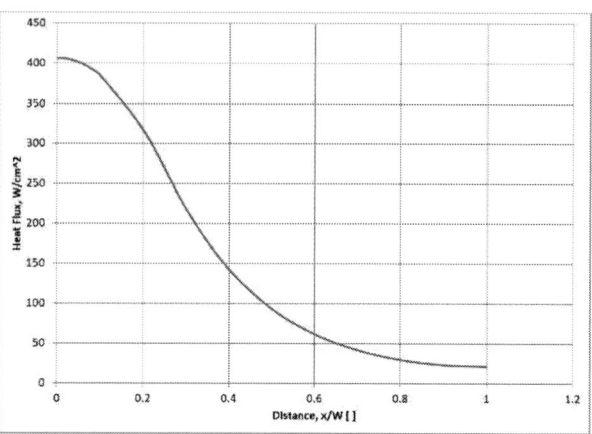

Figure 4 – Heat Flux distribution at fluid boundary for Rectangular sample problem (H = 3 mm)

4. ANALYTICAL SOLUTION – RADIAL GEOMETRY

In many applications the heat source and the heat spreader are arranged in more of square or circular configuration. In this case, a solution is needed for the hotspot temperature in a radial geometry. The assumption is made here that if the heat source is square or somewhat rectangular, it can be modeled as a circle with the equivalent area, i.e.

$$R = \sqrt{\frac{wL}{\pi}}$$

where w is the width of the component and L is its length. The configuration for the problem is shown in Figure 4.

Figure 4 – Geometry for the Radial Hotspot Analysis (left, side view; right, top view)

The differential equation to be solved is

$$\frac{\partial}{\partial r}\left(r\frac{\partial T}{\partial r}\right) + r\frac{\partial^2 T}{\partial y^2} = 0$$

$$\frac{\partial T(0,y)}{\partial r} = 0, \quad or \quad T(0,y) = finite$$

$$\frac{\partial T(R_o, y)}{\partial r} = 0$$

$$\frac{\partial T(r,0)}{\partial y} = \frac{h}{k}(T - T_o)$$

$$k\frac{\partial T(r,H)}{\partial y} = \frac{Q}{\pi R^2}f(r)$$

$$f(r) = 1, \; for \; 0 \le r \le R, and$$

$$f(r) = 0, \; for \; R < r \le R_o$$

As above, in the rectangular geometry, the method of separation of variables is used to solve this equation following the approach given in Arpaci [5]. The details of the solution are omitted here for the sake of brevity. The characteristic values, λ_n, are determined using the adiabatic boundary condition at R_o resulting in

$$\lambda_0 = 0 \; and \; \lambda_n \; are \; the \; roots \; of \; J_1(\lambda_n R_o) = 0 \; for \; n = 1,2,3, \ldots$$

Davis and Kirkham [6] provide the first 150 zeroes of $J_1(x)$ and Abramowitz and Stegun [7] provides McMahon's expansion for approximating the higher zeroes, in this case those above n=150.

The final solution for the temperature distribution is given below
$$\theta(r,y)$$
$$= q\left(\frac{R}{R_o}\right)^2\left\{\frac{1}{h} + \frac{y}{k}\right.$$
$$\left. + \frac{2}{k}\sum_{n=1}^{\infty}\left[\frac{\left(cosh\lambda_n y + \frac{h}{k\lambda_n}sinh\lambda_n y\right)}{\left(sinh\lambda_n H + \frac{h}{k\lambda_n}cosh\lambda_n H\right)}\frac{J_1(\lambda_n R)J_0(\lambda_n r)}{\lambda_n^2 R J_0^2(\lambda_n R_o)}\right]\right\}$$

The temperature at the hotspot is then given by

$$\theta(0,H) = q\left(\frac{R}{R_o}\right)^2\left\{\frac{1}{h} + \frac{H}{k}\right.$$
$$\left. + \frac{2}{k}\sum_{n=1}^{\infty}\left[\frac{\left(1 + \frac{h}{k\lambda_n}tanh\lambda_n H\right)}{\left(\frac{h}{k\lambda_n} + tanh\lambda_n H\right)}\frac{J_1(\lambda_n R)}{\lambda_n^2 R J_0^2(\lambda_n R_o)}\right]\right\}$$

And, the heat flux entering the heat sink at y=0 is given by

$$k\frac{\partial\theta(r,0)}{\partial y}$$
$$= q\left(\frac{R}{R_o}\right)^2\left\{1 + \frac{2h}{k}\sum_{n=1}^{\infty}\left[\frac{J_1(\lambda_n R)J_0(\lambda_n r)}{\lambda_n^2 R J_0^2(\lambda_n R_o)\left(sinh\lambda_n H + \frac{h}{k\lambda_n}cosh\lambda_n H\right)}\right]\right\}$$

5. RESULTS – RADIAL GEOMETRY

Table 2 shows a radial geometry sample problem similar to the rectangular geometry sample problem, described in Table 1. The heat source inputs 1,909 W/cm² of heat flux and the heat spreader wall rejects heat at 50,000 W/m²K in both analyses. The radius (r) of the disc-shaped heat spreader is 12.5 mm - same as half-length (W) of rectangular heat spreader. The thickness of the heat spreader wall is initially set at 1mm to likewise match Table 1.

Power of Heat Source =	60	W		
R = Radius of Heat Source =	0.00100	m	1.0	mm
Ro = Radius of Heat Spreader =	0.01250	m	12.5	mm
q = Heat Flux at Source =	1909.9	W/cm^2		
H =	0.001	m	1.00	mm
k =	400.00	W/m-K		
h =	50,000	W/m^2-K		W-in2-k
hH/k =	0.1250	[ ]	Biot number	
T$_{inf}$=T$_{v,OHP}$ =	0	C		
T$_{inf}$+q*(R/Ro)^2*(H/k+1/h) =	2.8	C		
T$_{inf}$+(R/Ro)^2*q/h =	2.4	C		Minimum Temperature
T$_{inf}$+q/h =	382.0	C		Maximum Temperature
T$_{hotspot}$ =	57.1	C		Hotspot Temperature

Table 2 – Radial Sample Problem Specifics

The increased heat spreader volume and surface area of the disc-shaped heat spreader reduced the hot spot temperature to 57°C above the fluid boundary temperature – an 80°C reduction from the rectangular heat spreader hot spot from Table 1. In addition, to the case in Table 2 with an h-value of 50,000 W/m²K, a quick analysis is able to be performed to see how a single-phase fluid heat rejection method effects performance by changing the h-value to 5,000 W/m²K – and results in a hotspot temperature of 91°C at 1mm wall thickness and a minimum hotspot of about 70°C at the optimized wall thickness of 3mm for this single-phase analysis.

Figure 6 shows the effect of disc thickness on the hotspot temperature for Table 2's sample case. The optimum thickness is about 4.4 mm. Recall that for the rectangular configuration the optimum is about 3.0 mm. This is reasonable since in the earlier case the heat load was relatively limited in surface to spread out in the direction into the page. In this radial layout, the heat flux is identical but the heat is able to conduct through a larger volume and heat rejection surface, if the wall thickness is large enough to allow for such spreading prior to reaching the two-phase boundary.

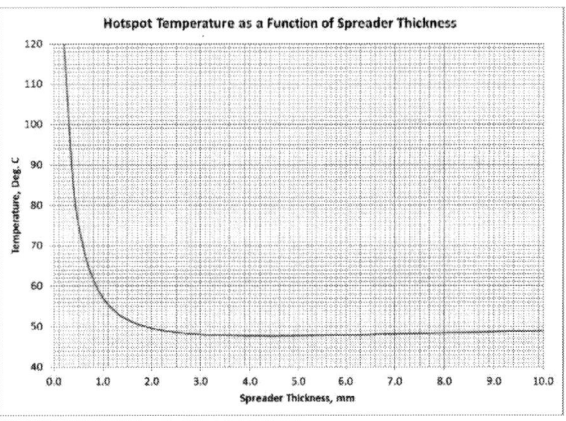

Figure 6 – Optimization Results for Radial Sample Problem

Figure 7 shows the predicted heat flux out from the radial heat spreader's wall (i.e., cooled surface or fluid boundary) at the optimal wall thickness – which in this case is just over 4mm. The peak heat flux has been transformed 1909 W/cm² to just under 35 W/cm². This is an order of magnitude lower heat flux into the fluid than in the rectangular case as shown in Figure 4 above.

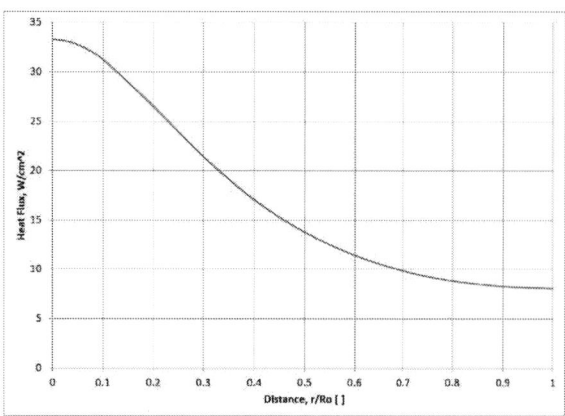

Figure 7 – Heat flux distribution at fluid boundary for Radial sample problem (H = 4 mm)

6. FINITE ELEMENT MODEL CORROBORATING RESULTS

The analytical model was validated against a finite element numerical model, using AutoDesk CFD® 2017 software package. For each configuration (rectangular and radial), there were four cases simulated – see Table 3. All other variables match those in Tables 1 and 2 for rectangular and radial configurations, respectively (i.e., 1,909 W/cm² heat source held constant, the Cu heat spreader thermal conductivity held constant, etc.).

Case	Film Coef. (W/m^2-K)	Wall Thickness (mm)
1	5,000	5
2	5,000	1
3	50,000	5
4	50,000	1

Table 3 – Four cases simulated with varying wall boundary conditions and thicknesses

A mesh analysis was conducted using the geometry and boundary conditions from Cases 1 and 2 as a baseline, by which to verify that a mesh independent solution was obtained. It was found that after the element count had reached approximately 50,000 elements per 1mm of thickness of material (all other dimensions held constant), that the solution was no longer dependent on the mesh element density. Therefore, the same approximate mesh density was used throughout the numerical analysis. Notably, for such high heat fluxes, nodal density is higher than required with lower flux applications.

Table 4 summarizes the results of the finite element analysis compared to the exact analytical solutions derived from the method presented above. Strong agreement exists between the two approaches; however, the analytical approach can be done in seconds using simple spreadsheet software while the numerical or simulation method requires relatively more expensive software and time to both build as well as iterate.

Case	Film Coefficient W/m²-K	Thickness mm	Maximum Temperature Analytical °C	Maximum Temperature Numerical °C	Maximum Heat Flux to Sink Analytical W/cm²	Maximum Heat Flux to Sink Numerical W/cm²
Rectangular 1	5,000	5	402.3	402.1	163.0	163.0
Rectangular 2	5,000	1	478.7	478.5	226.3	226.3
Rectangular 3	50,000	5	124.1	123.9	223.3	223.4
Rectangular 4	50,000	1	136.7	136.4	534.3	534.3
Radial 1	5,000	5	70.1	70.0	14.0	14.0
Radial 2	5,000	1	91.1	91.1	34.5	34.4
Radial 3	50,000	5	47.8	47.7	25.7	25.7
Radial 4	50,000	1	57.1	57.1	166.8	167.0

Table 4 – Summary of analytical results compared to numerical results (note: maximum temperatures are relative to a 0°C heat rejection boundary temperature)

Figures 8 and 9 present the eight simulations' outputs of the heat rejection fluxes to the heat sink (or fluid boundary) for the rectangular and radial configurations, respectively. Note, because this is an illustration of heat outflows from the surface of the heat spreader wall, the blue colors indicate where the most power is being rejected per unit area.

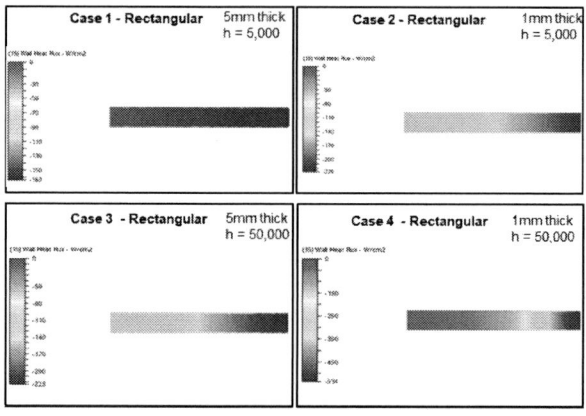

Figure 8 – Simulation images of heat flux output at fluid boundary for Rectangular cases

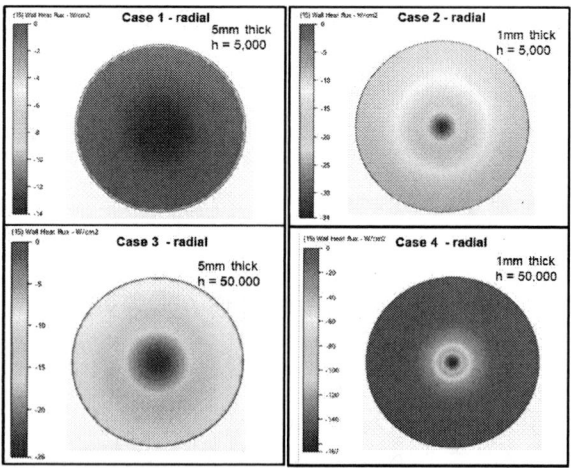

Figure 9 – Simulation images of heat flux output at fluid boundary for Radial cases

7. CONCLUSIONS

The analytical solutions presented clearly show that an optimum heated wall thickness exists and that arbitrarily decreasing the wall thickness of two-phase cooling solutions to minimize weight and decrease the resistance to the fluid does not yield the optimal solution for minimizing the high heat flux component temperature. The analytical approach can also be used to save time when initially scoping how to best thermally manage high heat flux devices and whether single-phase cooling or two-phase cooling is necessary to stay within specified operating temperature ranges. Further, the exact analytical solutions can allow engineers who use complex finite element simulations of real geometries a means to verify accuracy of such simulations' outputs.

ACKNOWLEDGMENTS

This material is based upon work supported by Contract No. N0017818C7000. Any opinions, findings and conclusions or recommendations expressed in this material are those of the author(s) and do not necessarily reflect the views of the Naval Sea Systems Command.

REFERENCES

1. Guo, H.A., Wiedenheft, K. and C-H Chen, "Hotspot Size Effect on Conductive Heat Spreading," IEEE Transactions on Components, Packaging and Manufacturing Technology, Vol. 7, No. 9, September 2017, pp. 1459-1464.

2. Lee, S., "Calculating Spreading Resistance in Heat Sinks," Design, Heat Sinks, Number 1, Test & Measurement, Volume 4, January 1998.

3. Gholami, A. and M. Bahrami, "Thermal Spreading Resistance Inside Anisotropic Plates with Arbitrarily Located Hotspots," Journal of Thermophysics and Heat Transfer, Vol. 28, No. 4, October–December 2014.

4. Thompson, S. M. and H. B. Ma, "Thermal Spreading Analysis of Rectangular Heat Spreader," J. Heat Transfer 136(6), pp. 064503-064511, Mar 11, 2014.

5. Arpaci, V.S., Conduction Heat Transfer, Addison Wesley, 1966, pg. 190.

6. Davis, H. T., and W. J. Kirkham, "A New Table of the Zeroes of the Bessel Functions J0(x) and J1(x) with Corresponding Values of J1(x) and J0(x)," Bull. Amer. Math. Soc., Volume 33, Number 6 (1927), 760-772.

7. M. Abramowitz & I. Stegun, Handbook of Mathematical Functions, Dover, New York, 1965, pg.371 (9.5.12).

Analysis of Natural Frequency Dependency on Temperature Variation of MEMS Vibratory Gyroscope

Jacek Nazdrowicz, Andrzej Napieralski

Lodz University of Technology, Department of Microelectronics and Computer Sciences

Wolczanska St. 221/223

Lodz, Poland

jnazdrowicz@dmcs.pl, napier@dmcs.pl

Abstract

In this paper we present method of analysis natural frequencies on temperature variation for one of the most popular motion sensor – gyroscope. This complex modeling, simulation and results analysis include variation of crucial quantities like: particular geometrical dimensions, physical properties of material, spring constants and damping coefficients for both drive and sense directions. Authors present theoretical background of modelling process and mathematical formulas governing gyroscope operation and points out quantities dependent on temperature. Simulations were performed in two stages with two software environments: COMSOL and Matlab/SIMULINK. Two different geometries were considered: with common mass for both resonator and accelerometer and with separate inertial mass for sense direction.

Results presented here and method of simulation can be very useful in further designing process of vibratory gyroscope operating in temperature variation environment (here between 293,15K and 393,15K) and improve performance by adjusting natural frequencies and Q factor.

Keywords

MEMS, vibratory gyroscope, spring constant, damping coefficient, natural frequency, temperature dependency.

Nomenclature

A   area, m^2

V   volume, m^3

a   source length, m

t   thickness, m

Q   quality factor

m   mass (kg)

T   temperature (K)

c   damping coefficient (Ns/m)

x   displacement in x (drive) direction (m)

y   displacement in y (sense) direction (m)

k   spring constant

v   linear velocity (m/s)

a   linear acceleration (m/s^2)

F   force (N)

d   inertial frame length dimension (m)

d   gap between movable part and substrate or cover (m)

Greek symbols

ρ   mass density (kg/m^3)

Ω   angular velocity (rad/s)

ω   frequency (rad/s)

μ   air viscosity (Pa·s)

Subscripts

x   x (drive direction)

y   y (sense direction)

d   external, drive

b   bottom (substrate)

u   upper (cover)

1. Introduction

Thermal simulations is essential step in analyzing microsystems like MEMS actuators or sensors. One of the main reason is that in most cases such microdevices operates in temperature dependent environment, what in case of micrometers scale structures, can influence meaningfully on performance and usefulness. In case of actuators – temperature variation is able to lead to instability of working; in case of sensor – errors in measurement. It just pays attention on fact that in electrostatic measurement method, capacitances (and more specifically – capacitance differences) are 10^{-14} F order what is very difficult to measure. Moreover initial vibrations caused by acceleration can have much larger amplitude, what is crucial to predict to avoid device destruction.

Problem of temperature influence on MEMS rotational sensors operation is not new. One can find some works including simulations regarding this topic. Paper [1] depicts method to simulate a MEMS vibratory gyroscope at various temperature. These simulations take into consideration two crucial quantities like Young's modulus and damping coefficient and results show temperature-dependent characteristics of the device. An effect of temperature influence on performance MEMS vibratory gyroscope is presented in work [2]. Here resonant frequency drop with the temperature increase is considered. Authors also discuss quality factor variation in dependency of temperature. In paper [3], the temperature model of a silicon gyroscope was considered and it was discussed Young's modulus deviation as temperature variation appears. In paper [4] temperature control system with PID algorithm was presented to improve the resonant frequency stability of a MEMS gyroscope [5].

2. Theoretical background

Basically, MEMS vibratory gyroscope is a damped mass on a spring. It contains a resonator (actuator, related to drive direction, here - x axis) and an accelerometer (sensor, related to sense direction, here - y axis), modeled as a two degree-of-freedom spring-mass-damper system. Principle of operation of vibratory gyroscope is not complex - it uses Coriolis effect (Coriolis force) closely related to rotating objects (fig.1 left).

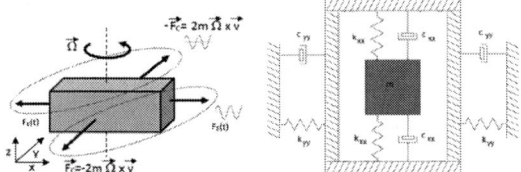

Fig. 1. General gyroscope and MEMS vibrating gyroscope operation principal (left) and general model of decoupled gyroscope (right).

Device rotating around z axis and vibrates along x axis with linear velocity v brings about vibrations along y axis (caused by Coriolis force). In case of clockwise rotation, Coriolis force acts along with negative part of y axis and in case of anticlockwise rotation, the force acts along with positive part of y axis.

Vibratory MEMS gyroscope can be depicted with second order differential equations in the following form [6]:

$$\begin{bmatrix} m_x & 0 \\ 0 & m_y \end{bmatrix} \begin{bmatrix} \frac{d^2x}{dt^2} \\ \frac{d^2y}{dt^2} \end{bmatrix} + \begin{bmatrix} c_x & 0 \\ 0 & c_y \end{bmatrix} \begin{bmatrix} \frac{dx}{dt} \\ \frac{dy}{dt} \end{bmatrix} + \begin{bmatrix} k_{xx} & k_{xy} \\ k_{yx} & k_{yy} \end{bmatrix} \begin{bmatrix} x \\ y \end{bmatrix} = \begin{bmatrix} -2m_x\Omega\frac{dy}{dt} + F_x \\ -2m_y\Omega\frac{dx}{dt} + F_y \end{bmatrix}$$

where k_{xx}, k_{yy} are spring constants for both modes (drive and sense respectively), F_x and F_y – external forces related to particular drive and sense directions, k_{xy} and k_{yx} are coupling stiffness coefficients of coupling between both modes (fig. 1 right). With assumption, that both modes are fully decoupled, system of equations can be simplified by conditions: k_{xy}, k_{yx}=0, F_y=0 and $2m\Omega\frac{dy}{dt} = 0$ and consequently above equations can be written in the following form:

$$m_x \frac{d^2x}{dt^2} + c_x \frac{dx}{dt} + k_x x = F_D \sin(\omega t) \quad (1)$$

$$m_y \frac{d^2y}{dt^2} + c_y \frac{dy}{dt} + k_y y = -2m_y \frac{dx}{dt} \Omega$$

or

$$\frac{d^2x}{dt^2} + \zeta_x \omega_x \frac{dx}{dt} + \omega_x^2 x = \frac{F_D}{m_x} \sin(\omega t) \quad (2)$$

$$\frac{d^2y}{dt^2} + \zeta_y \omega_y \frac{dy}{dt} + \omega_y^2 y = -2\frac{dx}{dt}\Omega$$

where $\zeta_x = \frac{c_x}{m_x\omega_x}$, $\zeta_y = \frac{c_y}{m_y\omega_y}$, ζ is a damping ratio, $\omega_x = \sqrt{\frac{k_x}{m_x}}$, $\omega_y = \sqrt{\frac{k_y}{m_y}}$, $Q_x = \frac{m_x\omega_x}{c_x}$, $Q_y = \frac{m_y\omega_y}{c_y}$ (3)

Note, that in a vast majority of simulations, geometrical dimension parameters and material physical properties are assumed to be temperature independent. However, this approach limits designed device to relatively constant operation temperature. In majority MEMS applications, it is required to analyze sensitivity of a vibrating gyroscope eigenfrequencies in relation to temperature fluctuation. Based on above considerations, MEMS vibrating gyroscope requires a frequency stability under changes taking place in the environment in which device operates. When we observe example results from simulations performed in Matlab/SIMULINK (fig.2) we can see that maximum

displacements for drive and sense directions are non-linear. As frequency grows, amplitudes decreases and for 36680Hz are almost 100 times less than for 9321Hz in case of x oscillations and almost 1000 times less in case of y oscillations. Results presented in fig. 3 show how in case of specified geometry maximum displacement in sense direction (in resonance) is sensitive in relation to applied frequency. Even small changes in frequency can cause meaningful changes in amplitude in sense direction. Additionally, geometrical dimensions can be substantial factor which decides of eigenfrequency value.

The most important for such kind of vibratory gyroscope is to obtain as maximum as possible displacement in sense direction. This is because non-zero Coriolis force gives very small amplitude in sensor as compared to resonator vibration. Therefore it is necessary to use resonance phenomena.

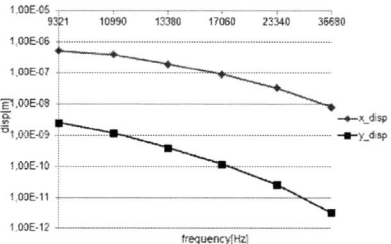

Fig. 2. Amplitude plots for drive and sense directions.

Based on above, resonance requires to apply external force with precise frequency. Even small deviation from the resonance value causes huge loses in vibration amplitude. Therefore it is incredibly crucial to consider in simulations all elements which can be influenced by temperature.

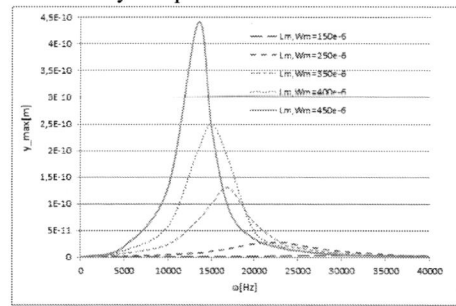

Fig. 3. Amplitude plots in frequency domain for different dimensions of spring constant.

After analysis of thermal expansion phenomena of gyroscope structure we specified components dependent on temperature. There are:

- geometrical dimensions:
$$l(T) = l_0(1 + \alpha\Delta T)$$
- Young's modulus [7]:
$$E(T) = E_0(1 + \beta\Delta T)$$
- spring constant:
$$k(T) = k_0(1 + \alpha\Delta T)$$
- damping coefficient [8-10]:
$$c(T) = \frac{\mu A(T)}{(1+2K_n)d(T)}, \quad K_n = \frac{\lambda}{L_c} = \frac{RT}{\sqrt{2}\pi D^2 N_a P L_c}$$
- thermal expansion coefficient $\alpha(T)$ [11-12].

where K_n is Knudsen number. Damping coefficient was used for Quality factor coefficient calculation which was then used to obtain magnitude and phase for resonator and accelerometer.

3. Models of vibratory gyroscopes

Modified models of MEMS vibrating gyroscope were prepared in COMSOL Multiphysics software. To simplify doing modification, models consist of fully-parametrized geometrical subparts. This allows to easy modify whole structure as well as particular elements: dimensions and locations. Simulation were done in two steps. First step was stationary solution to calculate thermal expansion of the device under applied temperature $T_0+\Delta T$ and second step was eigenfrequency analysis to calculate natural frequencies of particular modes and also visualize shape deformation of particular modes.

For simulation purposes physical parameters taken into account are presented in Table 1. In Table 2 list of crucial geometrical dimensions are shown for specified gyroscopes.

Symbol	Quantity	Value
μ	Air viscosity	$1.8 \cdot 10^{-5} \text{Ns/m}^2$
ε_o	Permittivity coefficient	$8.854 \cdot 10^{12} \text{F/m,}$
ρ	Density (Polysilicon)	2328 kg/m^3
α	Thermal expansion coefficient	$2.9 \cdot 10^{-5}$ 1/K
β	Thermal coefficient of Young's modulus β (for ΔT=0)	-80*10-6[1/K]

Table 1: List of physical properties of Polysilicon.

Quantity	Value
Drive, Sense electrode count	30
Distance between electrodes	10^{-5}m
Spring length	$250 \cdot 10^{-6}$m
Proof mass width	$1000 \cdot 10^{-6}$m
Proof mass length	$500, 1000 \cdot 10^{-6}$m
Device thickness	30^{-5}m
Central spring length	$140 \cdot 10^{-6}$m
Edge spring length	$250 \cdot 10^{-6}$m
Inertial frame height	$675 \cdot 10^{-6}$m

Table 2: List of some important geometrical parameters.

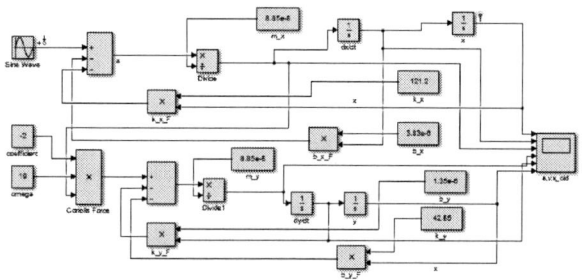

Fig. 4. Model of MEMS gyroscope in SIMULINK.

In fig. 4. there is model in SIMULINK presented which was also used in simulations to obtain maximum displacement (amplitude) in drive and sense directions.

4. Results of simulations and discussion

We started with base simulation which was taken for ΔT=0K: 293.15K. In fig. 5 there are presented results for

thermal expansion of device for 393.15K (ΔT=100K) with one mass for both directions, whereas in fig. 6 there are visualization expansion for device with one central mass and one inertial frame mass. We can observe that (according to our expectation), expansions take place in all directions – in other words, gyroscope expands its volume (mass does not change). In fig. 7 we see that along with distance from mass center total displacement grows, what confirms that particular parts change their physical dimensions. At the end of electrode expansion is about 0.1% for ΔT =100K. We observe also in fig. 5 and fig. 6 that springs deform meaningfully. It is very difficult to

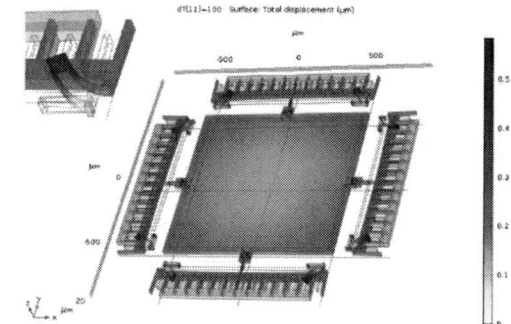

Fig. 5. Thermal expansion stationary analysis results for gyroscope with one inertial mass.

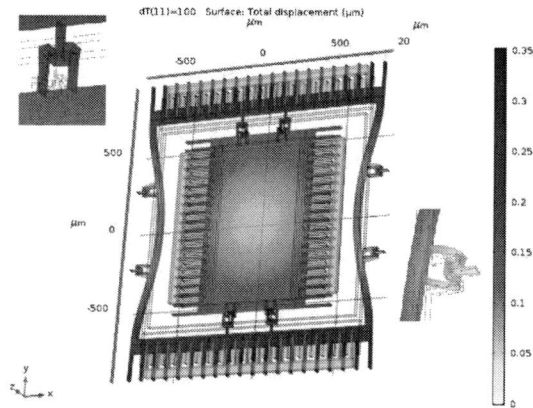

Fig. 6. Thermal expansion stationary analysis results for gyroscope with central inertial mass and inertial frame.

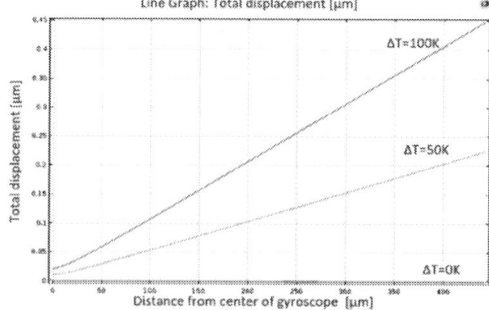

Fig. 7. Thermal expansion stationary analysis results.

calculate their spring constant with use known formulas for single beam spring constants. Therefore for further use spring constant is calculated based on results from Finite Element Method results. The fact of expansion and its magnitude is important because of two aspects: first is related to adjust

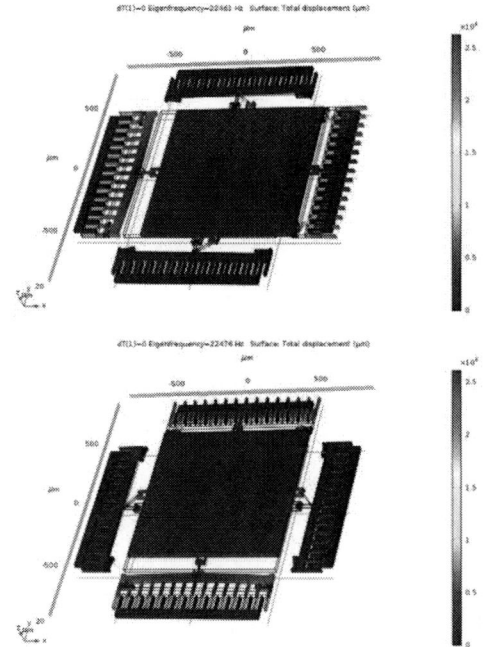

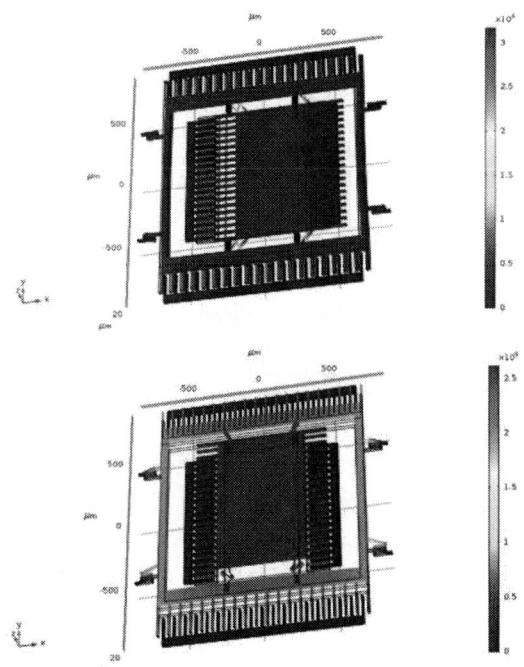

Fig. 8. First and second mode of MEMS gyroscope (drive direction) – for first geometry.

Fig. 9. First and second mode of MEMS gyroscope (drive direction) – for second geometry.

resonance frequency in sense direction and second is related to sensing process (electrodes expansion causes change of some geometrical quantities from capacitance equation).

In our considerations and simulations we were taken into account two modes reflecting vibrations along x (drive) and y (sense) directions. Deformation, total displacements and eigenfrequency (natural frequency) values for both gyroscope geometries are presented in fig. 8 and fig. 9. Depending on the inertial frame thickness and location of springs structure

expansion obviously influences on inertial frame deformation and causes additional stresses and, in turn, springiness.

For both gyroscope structures we see that selected modes are related strictly to particular drive and sense directions. We do not observe any undesirable deviation from these directions what can influence disadvantageously for gyroscope operation. Fig. 10 and 11 depict natural frequency dependency on temperature for gyroscopes with 1 mass and 2 masses configurations separately. As it is shown, in both cases eigenfrequency drops linearly as temperature grows. For gyroscope with 1 common mass (fig. 10) both plots are parallel whereas for gyroscope with inertial mass these plots have

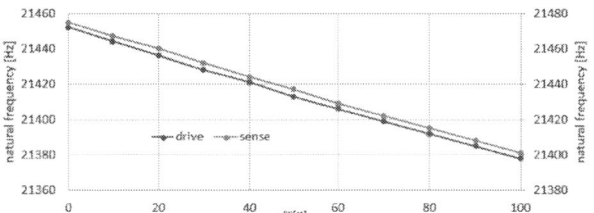

Fig. 10. Natural frequency dependency on temperature for gyroscope with 1 common mass.

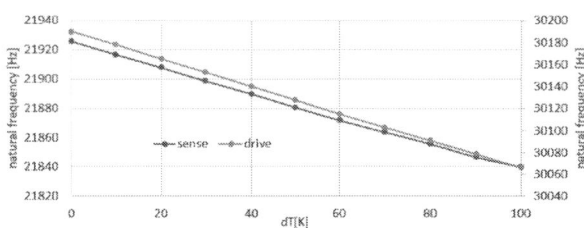

Fig. 11. Natural frequency dependency on temperature for gyroscope with 2 masses (including inertial mass.

different slopes - for drive direction frequency drops faster than for sense direction. This obviously has negative impact on frequency mode-matching.

In fig. 12-15 we can see results of simulations performed for different frame thickness, different temperature and different sense springs locations. Locations are marked as distance/4 (*d/4*), distance/3(*d/3*), distance/2(*d/2*) (1/4, 1/3 and 1/2 of frame length distance from its symmetry axis respectively). First of all, natural frequency for drive axis grows in whole range of inertial frame thickness independently on spring locations. However, in each case there is threshold of thickness below that natural frequency grows faster and above that grows slower. For sense direction, we observe that there is range of inertial frame thickness (below threshold) where natural frequency grows and range where natural frequency drops. It is particularly important fact because resonator

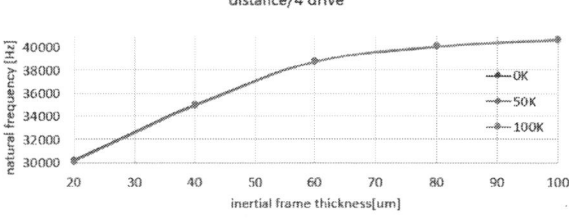

Fig. 12. Natural frequency dependency on inertial frame thickness for different temperatures and for spring locations 1/4 distance from symmetry axis (drive direction).

frequency should be adjusted to sensor natural frequency. As distance between springs locations grow, natural frequency dependency on thickness decreases (compare fig. 13 and 15).

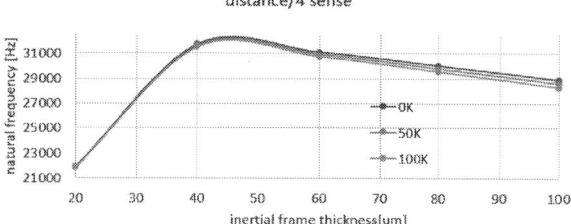

Fig. 13. Natural frequency dependency on inertial frame thickness for different temperatures and for spring locations 1/4 distance from symmetry axis (sense direction).

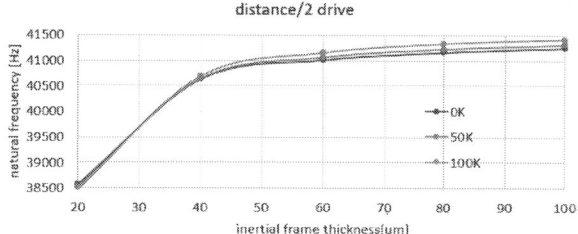

Fig. 14. Natural frequency dependency on inertial frame thickness for different temperatures and for spring locations 1/2 distance from symmetry axis (drive direction) - gyroscope with two masses.

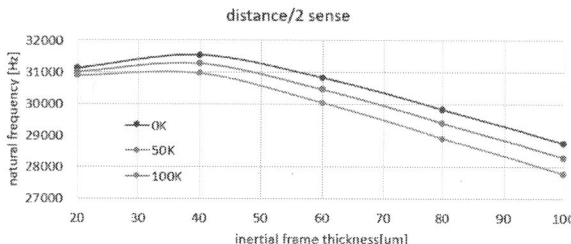

Fig. 15. Natural frequency dependency on inertial frame thickness for different temperatures and for spring locations 1/2 distance from symmetry axis (sense direction) - gyroscope with two masses.

For all springs locations eigenfrequency is lower for ΔT=100K than for ΔT=0K in whole range of thickness, and difference between these two values of temperature increases as thickness dimension increases.

Very interesting results are presented in fig. 16 and 17. These present natural frequency differences (between 0-50K and 50-100K) for inertial frame thickness for drive and sense separately. We can observe that for each plot there is minimum for given frame thickness. This minimum represents the smallest temperature-sensitive dependency for given inertial frame configuration and for given temperature range. Of course, the more important is sense direction therefore first, this one direction should be taken into consideration. For example, for temperature range 0-50K the most optimal should be configuration with thickness of 20 μm (the lowest value of natural frequency in sense direction) with $d/4$ distance of springs from symmetry axis.

Plots in fig. 16 and 17 confirm that there is strong dependency between natural frequency and temperature. Each modification of inertial frame dimensions causes meaningful changes in natural frequency. When temperature change –

frequency shift appears (for example for d/2 configuration – 40Hz shift) As expected, according to results presented in fig. 11-15 narrower range of eigenfrequency variation is for low temperatures than for higher ones for each suspensions locations and for both motion directions. For drive direction

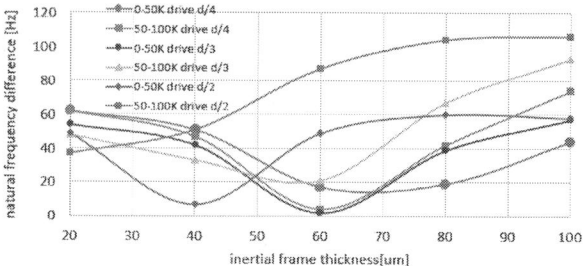

Fig. 16. Natural frequency dependency difference on inertial frame thickness for 0-50K and for 50-100K temperature difference – for all considered spring locations and drive direction - gyroscope with two masses.

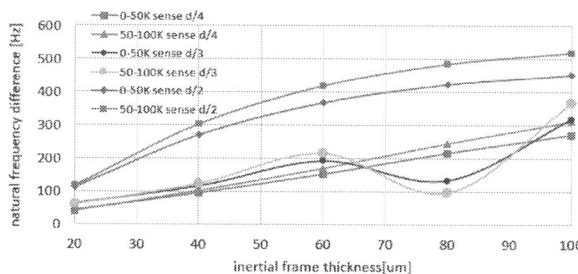

Fig. 17. Natural frequency dependency difference on inertial frame thickness for 0-50K and for 50-100K temperature difference – for all considered spring locations and sense direction - gyroscope with two masses.

these dependencies are different than for sense direction. For drive – natural frequency difference drops first and then increase for each spring configuration (fig. 16) whereas for sense direction in most cases it continuously grows (fig. 17). For 1/3 distance configuration there is exception – it drops for thickness of 60-80 μm in whole range of temperature. Looking at these plots it can be deduced what configuration is less and what configuration is more temperature sensitive for given

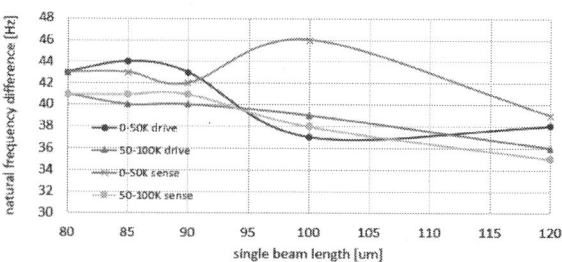

Fig. 18. Natural frequency dependency difference on central spring length single beam for 0-50K and for 50-100K temperature difference – for all considered spring locations and sense direction –gyroscope with one mass.

geometry configuration regarding natural frequency. Here for sense direction the best option is inertial fame with 20 micrometers with location d/4, because in whole range of temperature we can expect natural frequency variation 0 to

45Hz, what is very good results for configuration with 2-mass gyroscope. When we look at drive direction plots (fig. 13) we can expect 60Hz of eigenfrequency difference for this configuration (ΔT between 0 and 100K). Recall however, for thin frame additional factor to spring constant may need to be taken into account which potentially may be negative leverage for gyroscope operation.

In case of gyroscope with one mass tests were performed for different central mass dimensions (total length was constant, but lengths of particular beams changed). Results presented in fig. 18. shows strong dependency between natural frequency and temperature for drive and sense directions for ΔT between 0 and 50K. This dependency is stronger for long single beam (more than 90 µm).

The crucial for MEMS Gyroscope is that natural frequencies should be mode-matched for optimal performance. Because we do not know what is the optimal thickness, based

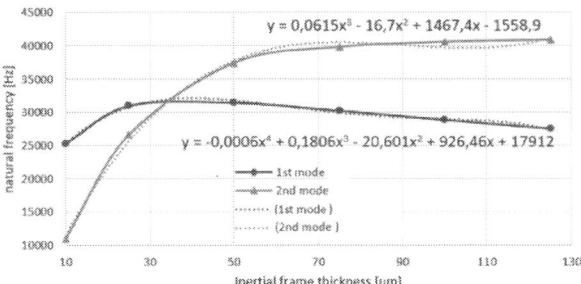

Fig. 19. Natural frequency dependency on temperature difference – for d/3 configuration and for drive (1st mode) and sense (2nd mode) directions.

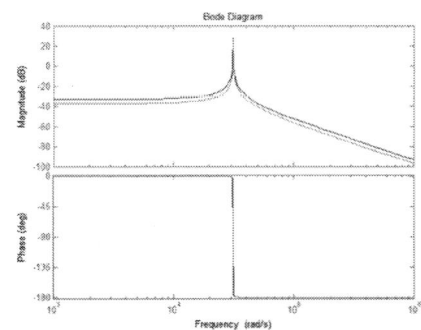

Fig. 20. Magnitude and phase – for d/3 configuration and for drive (1st mode) and sense (2nd mode) directions.

on plots presented in fig. 19 (plots for 1st and 2nd mode crosses for about thickness of 30 µm), it can be fit polynomials to these results obtained from simulations. For drive polynomial was 4th order, for sense – 3rd order. To obtain frequency value of mode-matched, system of both polynomial equations was solved with Matlab. It turned out, optimal thickness was 31,76µm. This value was used then in FEM simulation. All results are shown in table III. Damping coefficients calculated with SIMULINK model (based on area for given temperature) allowed us to obtain Q factors necessary to show magnitude and phase plots, presented in fig. 20.

5. Conclusions

In this article we presented analysis thermal influence on MEMS vibrating gyroscope eigenfrequency changes done with COMSOL Multiphysics software. Results taken from simulations show that problem of this quantity fluctuation is very serious and depends not only on temperature itself, but also on geometrical dimensions and some material properties like Young's modulus – which, in turn, also depends on temperature (decreases). In our simulations for 393.15K Young's Modulus decreased about 0,8% (for ΔT=0K, E=169GPa, for ΔT=100K, E=167,65GPa).

Results obtained for two different geometries confirmed that value of natural frequency is thermo-sensitive, moreover change of one geometrical dimensions require individual approach and perform series of simulations to verify if given configuration is optimal. Here, it is worth to underline, that even small natural frequencies (some Hz) can meaningfully impact on response of sensor (maximum displacement) causing mode-mismatching.

The less problem of mismatching is with configuration including one inertial mass, common for both drive and sense directions. Because natural frequency depends on mass and spring constant, it is enough to adjust spring constant parameter the same for resonator and accelerometer to obtain satisfactory results. In case of inertial frame this problem is more complex because for drive direction only central mass is taken into consideration, whereas for sense direction both masses (including inertial frame) need to be considered. However, as simulations shows (fig. 19 and 20), there is possible to find configuration for that drive and sense plots crosses and find common geometrical dimension that both natural frequencies are the same or very close (with use polynomial trend lines).

Results presented in paper confirm, that problem is multidimensional, analysis of temperature influence on whole device require individual approach to each structure. It is necessary to consider impact of dimension change of some crucial parts on total natural frequency and asses temperature influence on performance for given dimensions set.

This stage of design MEMS vibratory gyroscope is very important because it gives some hints regarding future device structure and its geometry details. Analysis natural frequencies comes from requirement to obtain as much as possible amplitude in accelerometer (y direction), because unexpected and unnecessary disturbances in case of such small mechanical devices are very likely and require compensation of signal to noise ratio by increasing mechanical output of accelerometer. This becomes crucial aspect of whole process leading to create device more robust and noises resistant.

Model of the gyroscope does not include any heat sources coming from adjacent electronic objects like IC, because this sensor is considered as separate device. Moreover it was assumed, that because gyroscope is very small device therefore in most cases temperature distribution is constant in its environment.

References

1. Chandradip P., P. MCCluskey, "Simulation of the MEMS Vibratory Gyroscope through Simulink", Conference: Device Packaging, March 2012, https://www.researchgate.net/publication/320170530_Simulation_of_the_MEMS_Vibratory_Gyroscope_through_Simulink

2. Xia D., S. Chen,S. Wang, H. Li, "Microgyroscope Temperature Effects and Compensation-Control Methods", Sensors, 9(10), pp. 8349-8376, 2009.

3. Fang J. C., J.L. Li, W. Sheng, "Improved temperature error model of silicon MEMS gyroscope with inside frame driving". J. Beijing Univ. Aeronaut. Astronaut. 32, pp.1277–1280, 2006.

4. A. Lawrence, "Modern Inertial Technology: Navigation, Guidance and Control", .Springer Verlag, New York,1993.

5. Guan R., C. He, D. Liu, Q. Zhao, Z. Yang, G. Yan, "A Temperature Control System Used for Improving Resonant Frequency Drift of MEMS Gyroscopes", Proceedings of the 10th IEEE International Conference on Nano/Micro Engineered and Molecular Systems (IEEE-NEMS 2015) Xi'an, China, April 7-11, pp. 397-400, 2015.

6. Jia J., X. Ding, Y. Gao, H. Li, Automatic Frequency Tuning Technology for Dual-Mass MEMS Gyroscope Based on a Quadrature Modulation Signal, Micromachines 2018, 9(10), https://www.mdpi.com/2072-666X/9/10/511

7. Yazdi N. et al., "Micromachined inertial sensors", Proceedings of the IEEE, vol. 86, pp. 1640-1659, 1998.

8. Cho Y.H., A.P. Pisano, R.T. Howe, "Viscous damping model laterally oscillating microstructures". Journal of Microelectromechanical Systems, 3, pp. 81–87, 1994.

9. Tang W.C., "Viscous air damping in laterally driven microresonators", In Proceedings of the IEEE Workshop on Micro Electro Mechanical Systems, Oiso, Japan, 25–28 January 1994, pp. 199–204, 1994.

10. Bao M., H. Yang, "Squeeze film air damping in MEMS", Sensors and Actuators A: Physical, 136, pp. 3–27, 2007.

11. Huang Q.-A., N.K.S. Lee, "Analytical modeling and optimization for a laterally driven polysilicon thermal actuator", Microsystem Technologies 5 (3), 133–137, 1999.

12. Paryab N., H. Jahed, A. Khajepour, "Creep and Fatigue Failure in Single- and Double Hot Arm MEMS Thermal Actuators", Journal of Failure Analysis and Prevention 9(2), pp.159-170, 2009.

Battery Discharge Capacity Calculation by Temperature Measurement

Jeevan Kanesalingam and Khoo Li Lian
Motorola Solutions
2A Medan Bayan Lepas,
Penang, Malaysia

jeevan.kanesalingam@motorolasolutions.com

Abstract

Battery capacity is measured in the unit of miliampere hour (mAh) (current x time). In order to do this, the discharge current of a battery is measured and integrated (multiplied) over the duration of the discharge. This paper proposes a method to calculate battery capacity by first measuring the temperature of a load resistor which is used to discharge the battery. The load resistor has a known/characterized Thermal Resistance (R_{th}) (degC.W^{-1}) value. The temperature is then integrated over time and solved using a thermal equation and thus obtaining the battery discharge capacity. The benefit of this type of discharge capacity calculation method is a series resistor (which causes a series voltage drop) is not required in the main current path.

Keywords

Capacity, battery, temperature

1. Introduction

Battery capacity is measured in the unit of mAh (current x time). The most common approach for a State of Charge (SOC) estimation is the Coulomb counting method. This is done by integrating the current (I) across time (t) [1], [2]. See Figure 1.

The approach proposed is to measure the temperature of an aluminum block to which power resistors have been attached to serve as a load to discharge the battery. Temperature of the block is measured to calculate the Power (W). The Power (W) is subsequently used to obtain the Current (I) which is then in turn used to calculate the capacity.

The measurement setup was done by using a custom built black aluminum block. Heater resistors were attached under the block to provide a heating source. A thermocouple was also placed inside the block at the center. Thermocouple readings of the block of aluminum + power resistors is initially characterized to measure the R_{th} value which is used in calculating the battery capacity.

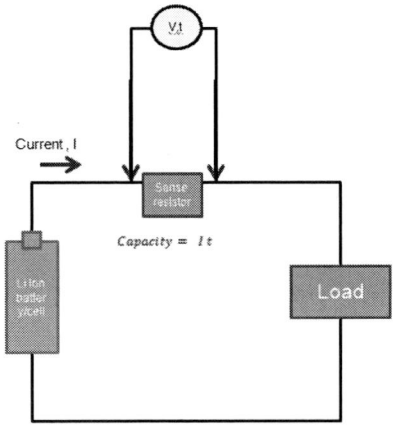

Figure 1: Typically setup for measuring battery discharge capacity

1.1. Calculation of capacity (mAh) using conventional method of current I

Capacity is calculated by I x t. For this to be done, a constant current needs to be used and the time t recorded. Alternatively the area under the curve from I vs. t graph can be used to calculate the capacity.

1.2. Calculation of capacity (mAh) using thermal measurements and R_{th}

1.2.1. Proposed setup for measuring battery discharge capacity by temperature calculation

As shown in Figure 2, proposed setup where temperature (T) and time (t) is recorded instead of current (I).

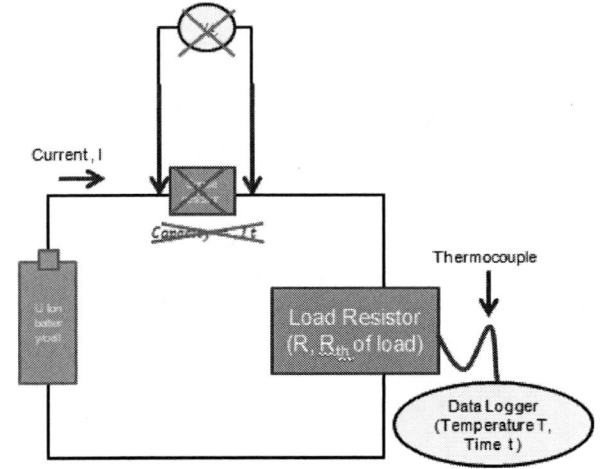

Figure 2: Proposed setup for measuring battery discharge capacity

1.2.2. Determining capacity (mAh) from Temperature (T) and Power (P)

The relationship between Temperature (T) and Power (P) is given as follows [3], [4]:-

$$T = P\,R_{th} + T_{amb} \tag{1}$$

Integrate (1) from 0 to t:

$$\int_0^t T = P\,R_{th}t + T_{amb}\,t + k \tag{2}$$

Let $k = 0$,

$$\int_0^t T = P\,R_{th}t + T_{amb}\,t \tag{2.1}$$

$$P = \frac{\int_0^t T - T_{amb}\,t}{R_{th}t} \tag{2.2}$$

$$P = I^2 R \tag{3}$$

$$I = \sqrt{\frac{P}{R}} \tag{3.1}$$

Substituting eq 2.2 into eq 3.1

$$I = \sqrt{\frac{\frac{\int_0^t T - T_{amb}\,t}{R_{th}t}}{R}} \tag{4}$$

$$Discharge\ Capacity = I.t \tag{5}$$

Substituting eq 4 into eq 5

$$Capacity = \sqrt{\frac{\frac{\int_0^t T - T_{amb}\,t}{R_{th}t}}{R}}\ .\ t \tag{6}$$

Where:-

$\Delta\,°C$ – Change in temperature (°C)

I – Calculated Current (A)

t – test time (s)

$\int_0^t T$ – Area under the curve for load resistor temperature across time (°C.s)

R_{th} – Discharge Resistor thermal resistance (°C/W)

P – Power dissipation across discharge resistor (W)

T – Temperature of resistor (°C)

T_{amb} – Ambient Temperature (°C)

1.2.3. Determining thermal resistance (R_{th})

Temperature of a surface can be calculated using the following equation:-

From equation (1),

$$R_{th} = \frac{T_j - T_{amb}}{P} \tag{7}$$

2. Equipment setup

2.1. Setup for thermal measurement with an anodized Aluminum block (Load Resistor)

Initially the aluminum block was setup per the following :-

- block needs to be attached with heater resistor (total resistance 3.33 ohm) to enable block to be heated up.
- heater resistors connected to a power supply
- a thermocouple should be attached inside the aluminum block to get the temperature right at the center of the block.

Figure 3: Side view of aluminum block with hole for insertion of thermocouple wire.

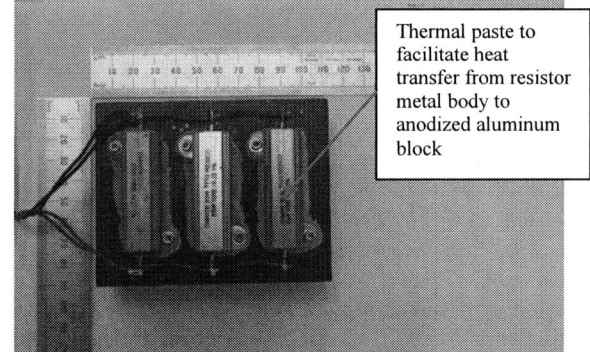

Thermal paste to facilitate heat transfer from resistor metal body to anodized aluminum block

Figure 4: Power resistors (3.33 ohms) to heat up Aluminum block

Equipment used:

Yokogawa DX2030 temperature reader with specs. :-

- Measurement temperature range (per calibration certificate) => -100°C to 800°C
- Reference junction compensation accuracy => ± 0.5 °C

Power to the Aluminum block with power resistors are supplied by black and red wires from a DC power supply. See Figure 4.

3. Test Method and Experiment Results

To obtain R_{th} and area under the curve for T vs. t, the following was conducted. A specific amount of capacity (I x t) was fed to the aluminum block and:-

- the temperature of the block was recorded across time T vs. t. (See Table 1 and Graph 1)
- R_{th} was calculated (See Table 1)

Vset	Discharge Current (A)	Power across load resistor, P (W)	Measured Temperature T(°C)	Room Temperature Tamb (°C)	(T - Tamb)/P = Rth (°C/W)
5	1.474	7.37	51.6	23.2	3.9
7	2.064	14.41	71.1	22.6	3.4
10	2.897	28.83	106.2	20	3.0

Table 1: Dissipated capacity across 3.33ohm and R_{th} calculation across the aluminum block

This value served and the check and balance value to compare against capacity calculation using the R_{th} method.

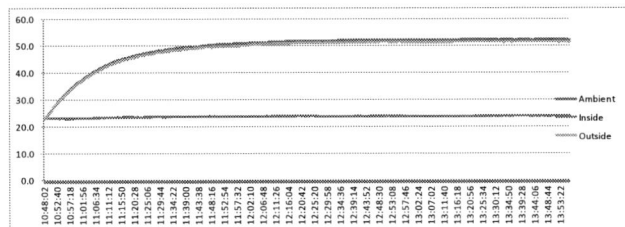

Graph 1: T vs. t curve @ 7.37W

Summing up the area under the curve for Graph 1, we obtain 154.5°C.s

Using eq. 5, $Discharge\ Capacity = I.t$

$$Discharge\ Capacity = 1.474 \times 3.17 \qquad (8)$$

$$Discharge\ Capacity = 4.669\ Ah \qquad (8.1)$$
(using conventional measured current, I multiply by time, t)

Substituting the following into (6),
$\int_0^t T = 154.5\ °C.s$,
$T_{amb} = 23.2\ °C$
$R_{th} = 3.9\ °C/W$
$t = 3.17$ hours
$R = 3.33$ ohms

$$Capacity = \sqrt{\dfrac{\dfrac{\int_0^t T - T_{amb}\ t}{R_{th}t}}{R}} \cdot t$$

$$Capacity = 4.427\ Ah$$

Comparing results obtained in eq 8.1 and 9, we observe a delta of only 5.3%

Repeating the above for different power levels, we obtain:

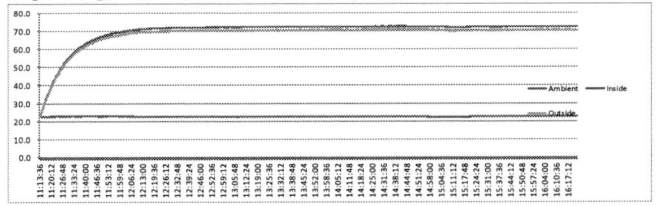

Graph 2: T vs. t curve @ 14.41W

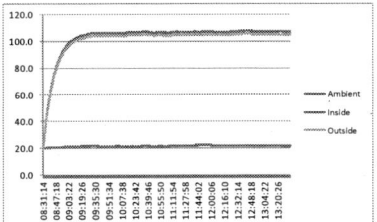

Graph 3: T vs. t curve @ 28.83W

Power across load resistor, P (W)	Test time, t (hours)	Using conventional capacity calculation method		I-calculated (A) using equation 4	Using temperature to calculate discharge capacity	% error from conventional method
		Discharge Current (A)	capacity I x t (Ah)		Discharge Capacity I x t (Ah)	
7.37	3.17	1.474	4.669	1.395	4.427	5.2
14.41	5.17	2.064	10.640	2.030	10.493	1.4
28.83	5.1	2.897	14.701	2.855	14.568	0.9

Table 2: Discharge capacity calculation comparison between conventional method and temperature measurement method

4. Measurement uncertainty discussion

The accuracy of the capacity calculation is heavily dependent on the accuracy of the measured aluminum block temperature and measured ambient temperature. The measurement uncertainty of the temperature was studied.

Symbol	Source of Uncertainty	Value	Unit	Distribution	Divisor for distribution	Uncertainty	Uncertainty
a	DX2030 Reference junction compensation accuracy	0.50	°C	R	1.73	0.29	0.0833
b	DX2030 Measurement Accuracy (digital display) @ 106°C	0.66	°C	R	1.73	0.38	0.1448
c	Thermocouple Tolerance °C	1.10	°C	R	1.73	0.64	0.4033
$U_{c\_temperature}$	Combined Uncertainty $\sqrt{U^2_a + U^2_b + U^2_c}$						0.7946
$U_{\_temperature}$	Expanded Uncertainty (95% confidence level, K=1.96)						1.5575

Table 3: Uncertainty Calculation for temperature measurements (using 106°C)

Table 3 shows that the measurement uncertainty is ±1.55°C at 106°C with a 95% Confidence Level. In % this is 1.46% of temperature uncertainty.

From eq 6, it can be observed that temperature accuracy has a direct impact on the final capacity measurement accuracy.

From eq 6, the relationship between I and T is as follows:
$$I \propto \sqrt{T} \qquad (9)$$

Substituting uncertainty of T 1.46% into eq 9, we obtain
$$I \propto \sqrt{1.46}\% \qquad (10)$$

With this the accuracy of the current calculation using eg 6 is ±1.21%

		Using conventional capacity calculation method			Using temperature to calculate discharge capacity		
Power across load resistor, P (W)	Test time, t (hours)	Discharge Current (A)	capacity I x t (Ah)		I-calculated (A) using equation 4	Discharge Capacity I x t (Ah)	% error from conventional method
7.37	3.17	1.474	4.669		1.412	4.476	4.1
14.41	5.17	2.064	10.640		2.054	10.620	0.2
28.83	5.1	2.897	14.701		2.890	14.739	-0.3

Table 4: Discharge capacity calculation comparison between conventional method and temperature measurement method by factoring in the calculated current with a +1.21% increase

From Table 4 it can be observed that when the +1.21% increase in current is included into the calculation, the difference between the measured current and calculated current using eg 6 decreases.

5. Summary of Results

Capacity calculated using the conventional method of tracking I and t and capacity calculated using the R_{th} method yield similar results. Capacity obtained has an error of 5.3% when compared against the conventional method. The sources of this variation are attributed to the temperature measurement accuracy. Taking this into account the capacity difference drops to 4.1%.

6. Conclusion and further work

R_{th} method of capacity calculation is an alternative way of capacity calculation. Instead of measuring current (I) and time (t) to calculate capacity, capacity can alternatively be calculated using Temperature (T) and t.

The 5.3%, 1.4% and 0.9% deviation between the conventional capacity calculation method and the proposed method by using thermal measurement is an area of further work. It should also be studied why there is a different in R_{th} when different power levels are applied to the Aluminum block.

References

[1]   K. Soon, C. Moo, Y. Chen, and Y. Hsieh, "Enhanced coulomb counting method for estimating state-of-charge and state-of-health of lithium-ion batteries," *Appl. Energy*, vol. 86, no. 9, pp. 1506–1511, 2009.

[2]   T. Takegami and T. Wada, "State-Of-Charge and Parameter Estimation of Lithium-Ion Battery Using Dual Adaptive Filter," pp. 1332–1337, 2017.

[3]   J. Kanesalingam and F. Kung, "Using SMT Chip Resistors Beyond Their Rated Thermal Specification," 2018.

[4]   Vishay Application Note, "Thermal Management in Surface-Mounted Resistor Applications Thermal Management in Surface-Mounted Resistor Applications," 2011. .

Exploring Heatpipe Configurations for Package On Package (PoP) Cooling

Dr. Sankarananda Basak
Intel Corporation, CCG (Client Computing Group)
2200 Mission College Blvd.
Santa Clara, USA
sankarananda.basak@intel.com

Ryota Watanabe
Lenovo (Japan) Ltd., Thermal and Performance Design,
Minatomirai Center Bldg. 21F, 3-6-1 Minatomirai,
Nishik-ku, Yokohama, Kanagawa 220-0012, Japan
rwatanabe2@lenovo.com

Abstract

Package on package (PoP) is commonly used in handheld devices to save board area. This configuration has the memory package sitting on top of the System on Chip (SOC) package. However, this package configuration poses challenge for thermal cooling of the SOC, as the memory package considerably increases the thermal resistance from the top of the SOC to the thermal solution over the package. In mobile systems, heat pipes are commonly used as thermal solution to transfer heat away from the SOC to remote heat spreaders or heat sinks. Heat pipes are known for their high effective thermal conductivity. Heat pipes are normally attached to the top of a package. Since there is additional thermal resistance on top of the SOC in a PoP configuration, we tried a novel idea to attach a heat pipe behind the PCB, below the package, and check the comparative heat transfer benefits. A single-sided PCB configuration is required for this thermal solution.

Keywords

PoP, PCB, TIM, Heat pipe

Nomenclature

SOC: System on Chip
CPU: central processing unit (subset unit of the SOC)
PCB: Printed Circuit Board
PoP: Package On Package
CFD: Computational Fluid Dynamics
TIM: Thermal Interface Material
T_j_max: Maximum SOC temperature
T_case: Maximum temperature at the rear case
T_glass: Maximum temperature at the top glass
OCA: Optical Clear Adhesive

1. Introduction

In handheld devices like tablets and detachables, it is important to reduce board area to maximize the plan area and volume for battery, to get decent battery life. For this purpose, it is a common to stack the memory package on top of the SOC package, in a typical Package on Package (PoP) configuration.

One downside of this configuration is that, the memory package creates additional thermal resistance in the heat transfer path from the top of the high power SOC. This may affect both steady state and transient thermal performance of a package.

Heat pipes are often used to carry the heat from the top of a package to heat spreaders in a handheld device. They are very effective in heat transfer due to inherent high effective thermal conductivity and low thermal resistance [1, 2, 3]. High effective thermal conductivity results in lower temperature gradient [4], and hence lower T_j_max.

Since PoP creates a significant heat transfer barrier to the top of the package, in this paper, we have explored a novel approach of PoP cooling by attaching a heat pipe below the PCB, behind the package. This approach can only work in single sided boards, which is quite common in handheld electronics. Attaching the heat pipe to the bottom side of a PCB has the additional benefit of allowing a larger area of thermal interface material between the PCB and the coldplate, located at the base of a heat pipe. Larger cross-sectional area of the TIM results in lower thermal resistance.

A detailed Computational Fluid Dynamics (CFD) analysis has been conducted, using a commonly used software in the electronics industry. The thermal model of a 12 inch detachable device with a PoP SOC on a single-sided PCB has been used in this study. Models have been created and simulations run with (1) the heat pipe attached to the top of the SOC PoP package for the baseline case. For the exploratory case, (2) the PCB has been flipped and the heat pipe attached to the back of the PCB, behind the SOC. That way, the heat spreader assembly for the entire system, stays the same. Please note that in this study, a CPU intensive software application has been chosen, which causes an intense localized hotspot in the SOC. This causes the sustained power in the device to be T_j_max limited. (For an applied SOC and system power, T_j_max reaches its threshold temperature before T_case or T_glass reaches their respective threshold temperatures).

Best practices developed within the companies, from correlation of experimental and modeling results, has been used to develop the thermal models. Combination of correlated and published heat pipe thermal properties have been used to model the heat pipe.

2. Thermal models

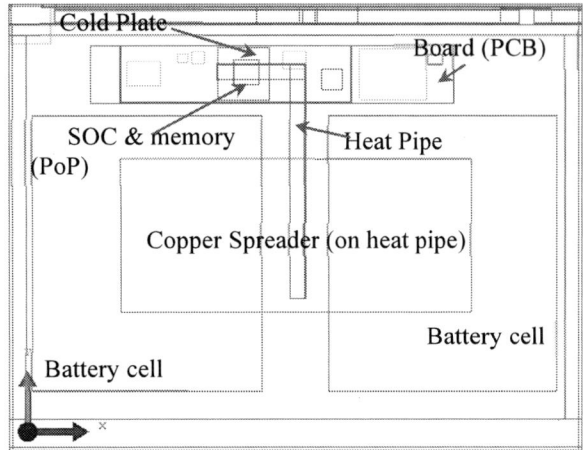

Figure 1: Plan view of thermal model of a 12 in detachable device in a CFD software.

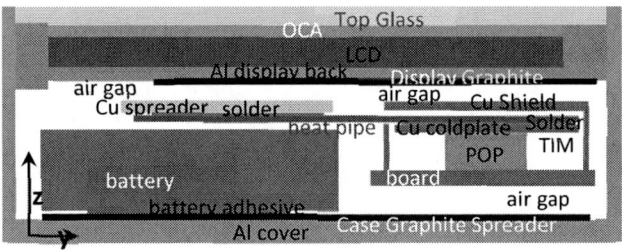

Figure 2: Side view of configuration 1. Standard heat pipe solution, with the heat pipe on top of the SOC package (PoP).

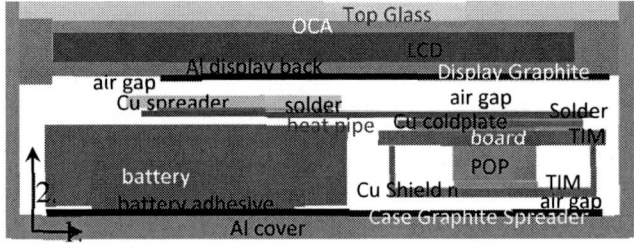

Figure 3: Side view of configuration 2. PCB flipped. Heat pipe thermally connected to the back of the PCB, behind the SOC.

The plan view of the 12 in detachable thermal model is shown in Figure 1. Different important components of the thermal model, like PCB, SOC and memory (PoP), heat pipe and heat spreaders, battery cells, are marked. In this model, x direction corresponds to the length of the detachable device, y corresponds to the width and z corresponds to the height.

The vertical stackups for configurations 1 and 2 are shown in Figures 2 and 3 respectively. The stackups through the SOC is also described in Table. 1.

Configuration 1		Configuration 2
Top Glass		Top Glass
OCA		OCA
LCD Panel		LCD Panel
Al case (behind LCD)		Al case (behind LCD)
Graphite layer		Graphite layer
Air Gap		Air Gap
Shield Can		Heat Pipe
Air Gap		Solder
Heat Pipe		Copper Cold Plate
Solder		TIM
Copper Cold Plate		Board
TIM		SOC, Memory PoP
SOC, Memory PoP		TIM
Board		Shield Can
Air gap		Air gap
Case graphite spreader		Case graphite spreader
Al rear case		Al rear case

Table 1: Vertical stackup through the SOC for configurations 1 and 2.

Configuration 1 (Figure 2) is the baseline case, showing a standard heat pipe configuration, attached to the top of the PoP. In this configuration, there is a hole in the shield can, through which the heat pipe exits and connects to the copper spreader on the left. A thin thermal interface material (TIM) is modeled between the cold plate (at the base of the heat pipe) and the top of the package. This interface is expected to be pressure mounted with screws, springs, etc. The area of the TIM is assumed to be the same as the plan area of the PoP.

Configuration 2 (Figure 3) is the exploratory configuration, where the PCB is flipped downward, and the heat pipe is attached on the opposite side of the PCB, behind the PoP. A thin thermal interface material is modeled between the back of the PCB and the cold plate for configuration 2. Due to the availability of a larger interfacial area between the board and the cold plate in configuration 2, a larger TIM area has been used (compared to configuration 1). Also, an additional gap filler is used between the package and the shield can in this case, thereby allowing some heat flow from the top of the PoP to the shield can as well. The shield can is continuous in configuration 2.

An ambient temperature of 25 C has been used in the studies.

Component	Thermal Conductivity K or K_x, K_y, K_z (W/mK)
Glass	1.1
OCA	0.2
LCD panel (bulk)	1.05
Al sheet (Display)	167.0
Graphite Spreader	450.0, 450.0, 1.0
Copper	385
TIM (base of heat pipe copper plate)	3.8
Solder (between Heat Pipe, copper plate)	30.0
Heat pipe vapor	50000.0
PCB	60, 50, 1.1
Al Case	167.0

Table 2: Thermal conductivities of some materials provided.

Thermal conductivities of some materials used in the models have been provided in Table. 2. Only thermal conductivity is relevant for steady state modeling and simulation, as has been explored in this paper. The PCB conductivity used is commonly observed in tablet or detachable form factors, especially in the highly routed region below the SOC.

Configuration 1	Configuration 2
12 mm x 12 mm x 0.05 mm	18 mm x 18 mm x 0.05 mm

Table 3: TIM dimension at the base of heat pipe copper cold plate.

TIM dimensions used for the two models, at the base of heat pipe copper plate is given in Table. 3. The heat pipe TIM area for configuration 2 is behind the PCB. It is not constrained by the SOC package area, and can be bigger. Hence a bigger TIM area has been used in this case to decrease the thermal resistance.

The heat pipe is modeled as 1 mm thick and 7.5 mm wide. It is representative of a 5 mm heat pipe crushed to 1mm thickness. There is outer copper thickness of 0.2 mm. Additional resistances in the evaporator region of 1.25e-4 (m^2K/W) has been applied. The thermal conductivity in the vapor region of the heat pipe is shown in Table. 2.

3. Results

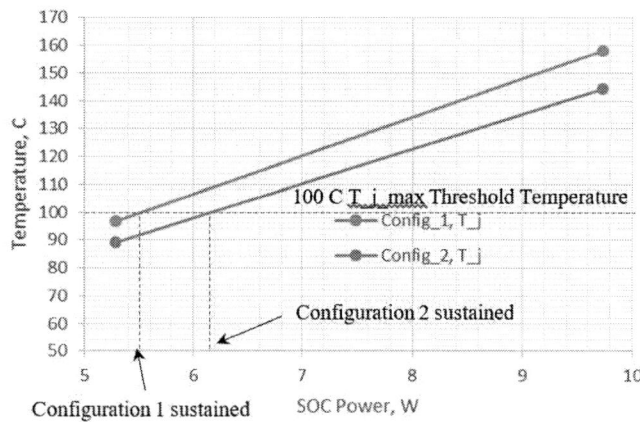

Figure 4: Variation of T_j_max vs SOC power for the 2 configurations studied.

Both the system thermal models, corresponding to configurations 1 and 2 has been simulated with 2 SOC power levels. The maximum SOC temperature (T_j_max) for the two SOC power levels for the 2 configurations have been shown above. T_j_max for configuration 2, the case with the heat pipe attached behind the PCB, is ~ 8%-9% lower at these power levels.

	Config: 1	Config: 2	Power % change
Sustained SOC power, W, (100 C, T_j_max limit)	5.5	6.2	11.6%

Table 4: Sustained SOC power, using the criterions of T_j_max=100 C. Configuration 2 allows ~ 11.6% more sustained SOC power for the system within the temperature limits.

Using a criterion that the SOC T_j_max is limited to 100 C, the sustained SOC power for configuration 1 is 5.5 W, while the same for configuration 2 is 6.2 W (Table 4). Configuration 2 is running cooler for the same SOC power, and hence, it can sustain larger SOC power (by ~11.6%) for the same threshold T_j_max.

Heat Split	Config: 1	Config: 2
Heat Flow package top %	57%	29%
Heat flow package base %	43%	71%

Table 5: Variation of heat split to the top and bottom of the package. Configuration 2 allows a lot more heat to flow to the bottom of the package.

Table. 5 shows the heat split for configurations 1 and 2. For configuration 1, with the heat pipe at the top of the PoP, ~ 57% of the PoP heat is going to the top, through the memory package. While for configuration 2, with the heat pipe attached on the opposite side of the board, behind the PoP, ~ 71% of the PoP heat is going through the PCB. The novel idea explored in configuration 2 is quite effective in transferring heat away from the SOC. This explains the overall SOC running cooler for configuration 2 for a given SOC power.

4. Conclusions

A novel system thermal solution, with heat pipe at the bottom of a PCB, behind the SOC for a PoP configuration (configuration 2) has been explored, and it has been compared to a more standard thermal solution with heat pipe attached to the top of the package (configuration 1). It is seen with the model configurations and assumptions, a heat pipe at the bottom of a PCB for a PoP package is more effective in heat transfer than heat pipe at the top. This is causing the SOC T_j_max to be lower at a given SOC power level. Consequently, this is allowing more SOC power to be sustained by the system within the same T_j_max criterion.

Acknowledgments

Contributions from Intel team-members Hua Zhang, Joshua Een, Nicholas Weber and Aastha Uppal are gratefully acknowledged.

References

1. Brennan, P.J. and Kroliczek, E.J., Heat PipeDesign Handbook, B&K Engineering, NASA Contract No. NAS5-23406, June 1979.
2. 2. Chi, S.W., Heat Pipe Theory and Practice, Hemisphere PublishingCorporation, 1976.
3. 5. Peterson, G.P., An Introduction to Heat Pipes Modeling, Testing, andApplications, John Wiley and Sons, Inc., 1994.
4. Incropera, Frank P., DeWitt David P, Fundamentals of Heat and Mass Transfer, Wiley (2002).

Simulation-Based Optimization of Data Center Cooling Performance Using Performance Indicators

John Petrongolo[1,3], Kourosh Nemati[2], and Kamran Fouladi[1]

[1]Widener University
1 University Place
Chester, PA 19013

[2]Future Facilities
2055 Gateway Pl, #110
San Jose, CA 95110

[3]Corresponding Author: japetrongolo@widener.edu

Abstract

This study investigates the effects of computer room air handler (CRAH) setpoints, room configurations, and containment strategies on a new set of metrics. The new metrics proposed by The Green Grid is called the performance indicator (PI) and include power usage effectiveness ratio (PUEr), thermal conformance, and thermal resilience. The study was conducted using the computation fluid dynamics software 6SigmaRoom®, and the results show substantial effects on thermal conformance and thermal resilience when the configurations are changed, and when containment strategies are utilized. The CRAH setpoint influences the PUEr the most with higher setpoints resulting in lower PUEr values.

Nomenclature

PUE	=	Power Usage Effectiveness
PUEr	=	Power Usage Effectiveness ratio
PI	=	Performance Indicator
CRAH	=	Computer room air handler
CFD	=	Computation fluid dynamics
SAT	=	Supply air temperature
COP	=	Coefficient of performance
RHI	=	Return heat index
SHI	=	Supply heat index

1. Introduction

Data centers are mission critical facilities that house vital IT equipment with cooling infrastructures designed to maintain the environmental conditions maximizing IT equipment reliability. Data centers are significant consumers of energy, and they have consistently increased their power usage in the past with no slowdown in sight. Reports have indicated that data center power consumption in the United States has grown drastically since the year 2000 [1]. A major operational goal for data center designers and managers is to have highly efficient data centers while keeping the IT equipment within a specific temperature range and performance metrics are useful in helping them accomplish this goal. There have been number of performance metrics proposed and used over the years, but Reddy et al. [2] argued that there is a need for more versatile and comprehensive metrics. They reasoned that the current metrics are incomplete, and a new set of metrics need to be developed that are more compatible with all types of data centers. For example, older data centers may not be able to use certain metrics because of the complexity of their measurements. Conversely, new metrics that can factor in different locations,

as well as the age of the data center, can allow for more comparisons between different data centers [2]. Within the past year, The Green Grid has set out to develop a new set of metrics, called the performance indicator (PI), that have the potential to predict the best future arrangements within data center [3]. The PI is made up of three distinct metrics, power usage effectiveness ratio (PUEr), thermal conformance, and thermal resilience. Each of the three metrics serve a specific purpose, but all three are interrelated. This means that a change to one metric can influence the other two. PUEr is based on the widely known metric PUE and it has been defined to better represent the efficiency goal of the data center using

$$PUEr = \frac{PUE_{reference}(X)}{PUE_{actaul}}$$
Equation 1

where $PUE_{reference}(X)$ is the lowest end of the aspired PUE set by the manager and PUE_{actaul} is the actual PUE of the data center. Thermal conformance represents the amount of IT equipment within the recommended inlet temperature range set by the American Society of heating, refrigeration, and heating engineers (ASHRAE). Thermal resilience is the value that represents the amount of IT equipment within the allowable inlet temperature range when there is equipment failure or scheduled repairs [3]. Thermal conformance and thermal resilience metrics are defined by Equation 2 and Equation 3, respectively.

$$IT\ Thermal\ Conformance$$
$$= \frac{Equipment\ Load[T_{\max inlet} < 27°C\ under\ normal\ operating\ condition]}{Total\ Equipment\ Load}$$
Equation 2

$$IT\ Thermal\ Resiliance$$
$$= \frac{Equipment\ Load[T_{\max inlet} < 32°C\ under\ worst\ case\ ACU\ failure]}{Total\ Equipment\ Load}$$
Equation 3

It is difficult to optimize all three metrics at the same time since they are interrelated. Therefore, there are tradeoffs that need to be considered based on the purpose of the data center. For instance, a data center that contains servers that always needs to be operational, such as stock trading servers, would need to prioritize thermal conformance, which may result in a decrease of PUEr. Conversely, a data center with search engine servers that are not critical would perhaps prioritize

PUEr to decrease their operational cost. This may cause thermal conformance or thermal resilience to decrease.

The present effort uses a simulation-based approach utilizing computational fluid dynamics (CFD) to investigate the effects of several data center parameters on the performance indicator. CFD has evolved into a vital tool for data center design and research and has helped to make important advances in the industry. For example, a number of recent efforts have used CFD to research the effects of parameters such as CRAH unit setpoints [4], airflow uniformity [5], room configurations [5], etc.

CRAH unit setpoints have been a major part of the recent research since it has a direct effect on the room's operating power. More specifically, a higher setpoint requires less power to operate, which results in lower costs. Therefore, extensive research into this parameter can help to lower operating costs because 30% -50% of power is spent on cooling the IT equipment [1]. Ham et al. conducted a study on setpoints with different types of cooling methods. Their CFD-based study resulted in general guidelines to have a supply air temperature (SAT) of 18°C to 23°C to improve efficiency [4].

The room configuration also has significant effects on the data center performance. For example, a less than ideal configuration can influence the premature mixing of hot and cold air, called recirculation, resulting in inefficiencies. S.A. Nada used numerical investigation to experiment with different CRAH unit configuration placement, and the results of the study showed a more uniform rack inlet temperature when there were CRAH units on either side of a row instead of solely on one side [5].

Containment for a data center involves containing the air either in the hot or the cold aisle. A cold aisle containment encases the cold aisles shown below in Figure 1, while the hot aisle containment encases the hot aisle and utilizes a false ceiling to cycle the air shown in Figure 2. There are many branches to both these strategies that involve partial containment, but the originals are the most influential.

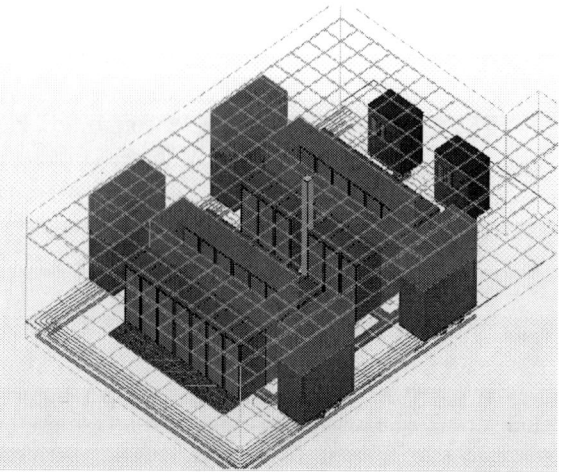

Figure 2. Representive of hot aisle containment.

There are some advantages and drawbacks to each strategy, but the containment strategies have shown to improve the efficiency of the cooling system [6]. It has been argued that the cold aisle strategy is not as beneficial as hot aisle, but it is often implemented in legacy data centers due to its easier construction and lower cost [6]. The hot aisle containment is overall a better strategy by allowing the ability to set the work environment temperature and enabling the economizer to run longer hours throughout the year [6]. The hot aisle containment is shown to result in a savings of 43% in annual cooling system energy cost when compared to cold aisle containment [6].

This paper is focused on investigating the metrics within the PI and how they affect each other when an aspect of the data center is changed. The data center parameters of interest include CRAH setpoint, room configuration, and containment strategy and the effect of each parameter were studied independently. The parameters chosen are considered influential in a data center's efficiency, but little is known about how they interact with the PI [3]. The numerical simulations in this study are performed using the commercially available numerical code 6SigmaRoom® [7].

2. Numerical Methods

The model, shown in Figure 3, contains 12 CRAHs, 180 PDUs, and 1,412 IT equipment scattered across 12 different rows. The aisles are arranged in a hot/cold aisle configuration and have blanking tiles on racks where there is no IT equipment.

Figure 3. Isometric view of data center

The CRAH units utilize a chilled water system as the cooling method. A coefficient of performance (COP) is associated with the chilled water system to calculate the amount of energy it takes to cool the water to its intended

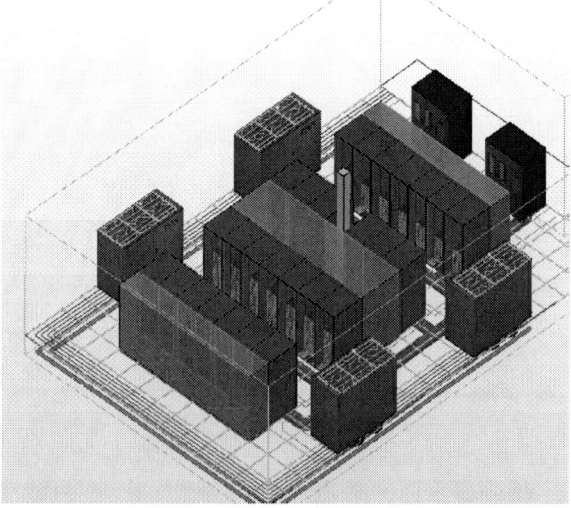

Figure 1. Representive of cold aisle containment.

setpoint. The COP that varies by the chilled water setpoint is defined by Equation 4, where T_{supply} represents the chiller setpoint [8]. This relation then enables PUE to change when the setpoint of the chilled water changes.

$$COP = 0.0068T_{supply}^2 + 0.0008T_{supply} + 0.458$$
Equation 4

Each CRAH units in the above configuration is set at its own setpoint, which results in no universal value to change for each trial. Therefore, the chilled water system is used as the variable in the study since the CRAH setpoints are tied to the chilled water system setpoint and would change as the temperature of the chilled water system changes. The chilled water is then varied from 8.33°C to 13.33°C, and the CRAH unit setpoints are then changed accordingly. The setpoint study configuration is shown on the right in Figure 4. This figure displays the power of each cabinet throughout the data center.

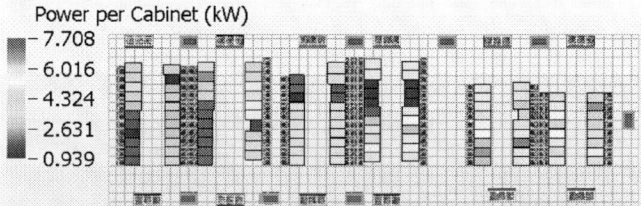

Figure 4. Power distribution during setpoint study

The room configuration study was conducted on eight different configurations. Table 1 gives a brief description of the configurations. These trials were conducted on higher density configurations shown in Figure 5. The higher density allows for more drastic changes in the thermal conformance and the thermal resilience. The configuration trials consisted of relocating cabinets, IT equipment, and rows in the data center. Table 1 gives a brief description of the configurations. Configuration A is the original while B is a slight relocation of the cabinets to centralize them through the data center. The higher density cabinets as well as cabinets requiring higher airflow were moved to centralized locations. These configurations are represented by letters C through F in Table 1. The high-density area shown on the far right in Figure 5 was distributed around the data center in configuration G. Configuration H has IT equipment rearranged to cabinets with more available air. Finally, the IT equipment switches, which are normally placed in the back of cabinets, were moved to the front of the cabinets in configuration I.

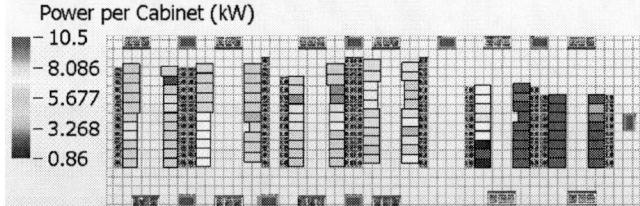

Figure 5. Power distribution during configuration and containment study

A	Original Configuration
B	Rows are centralized within the room
C	Rows are centered within the room according to power
D	Rows are centered within the room according to the power and equally distributed
E	All rows are centered within the room according to airflow
F	all rows are centered within the room according to airflow and equally distributed
G	Power density is spread out and rows are centered for airflow
H	IT equipment is rearranged according to excess airflow
I	IT equipment is rearranged, and IT switches are brought to the front of the cabinets

Table 1. Descriptions of configurations

The configuration study results were analyzed and then used with the containment study. The most optimal configuration and one of the least optimal configurations were used in the containment study. The most optimal configuration was configuration I when the IT equipment was rearranged, and switches were moved to the front of the cabinet. The other configuration chosen was configuration D, the power-heavy cabinets located toward the center of the aisle and all rows distributed evenly across the data center. There were three containment strategies used: partial containment hot aisle, full containment hot aisle, and cold aisle containment. The partial containment involved enclosing the hot aisle, but the roof of the containment had holes in it for hot air to escape shown in Figure 6. The full containment for the hot aisle utilized a false ceiling where the hot air would return to the CRAHs. The cold aisle containment was used to encompass the cold air before it ran through the cabinets.

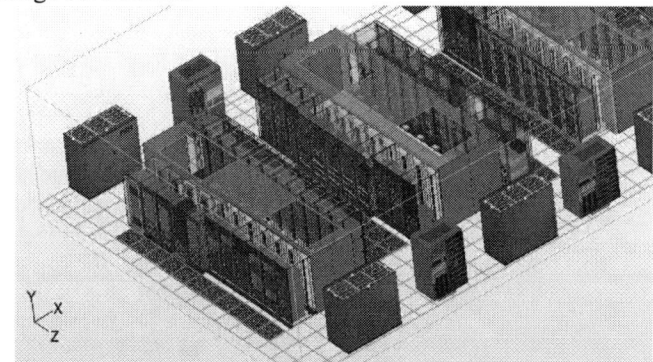

Figure 6. Represenation of partial hot aisle containment

3. Results

3.1. Results of Setpoint

The results for the CRAH units setpoint study are shown below in Figure 7 and Figure 8. Figure 7 displays a linear increase in PUEr with each temperature change. Figure 8 combines both the results for thermal conformance and thermal resilience. For thermal resilience, the results indicate a minimal change in the values, whereas there is a notable change in the thermal conformance.

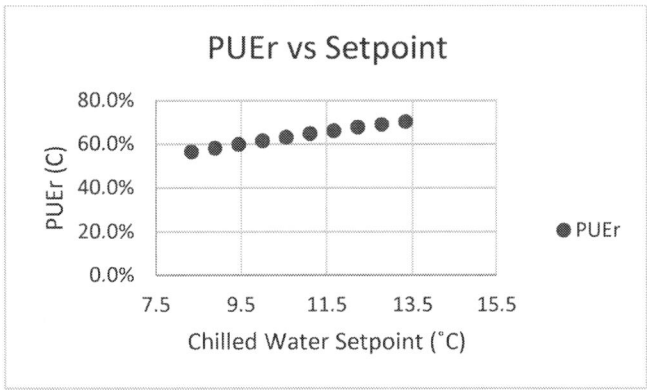

Figure 7. Results of PUEr for setpoint study

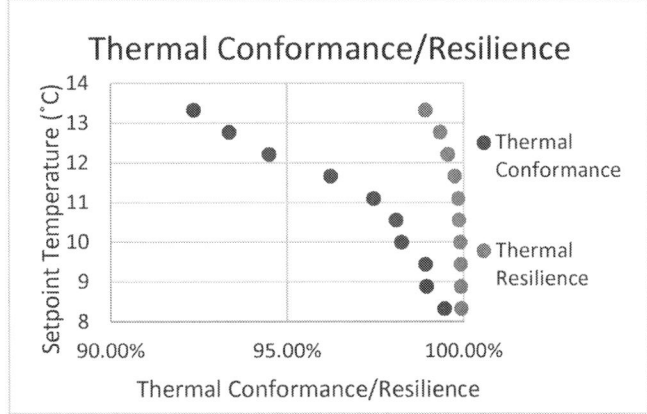

Figure 8. Results of thermal conformance/resilience for setpoint study

3.2. Results of Configuration

The configuration changes in the study showed a much greater effect on the thermal conformance and thermal resilience due to the change in power density, as discussed earlier. The configurations focused on the placement of cabinets with higher airflow or higher power requirement in the center showed positive effects on the thermal conformance and the thermal resilience. The configurations that distributed the distance between rows had adverse effects on the PI metrics. This may be due to the fact that the gap between the rows was reduced allowing for more mixing of hot and cold air. Significant benefits were observed when the IT equipment was rearranged (Configuration I). In this configuration, the IT equipment were moved to cabinets that had greater excess airflow, which increased the thermal conformance by almost 5% compared to the original configuration. Subsequently, the IT equipment switches were then rearranged. These switches are typically placed in the back of the cabinets but were moved to the front of the cabinet for this particular configuration (Configuration I). This rearrangement resulted in 10% increase in thermal conformance compared to the original configuration. Figure 9 shows the results of the configuration study and it is evident that not all configurations had significant impact on the PI metrics.

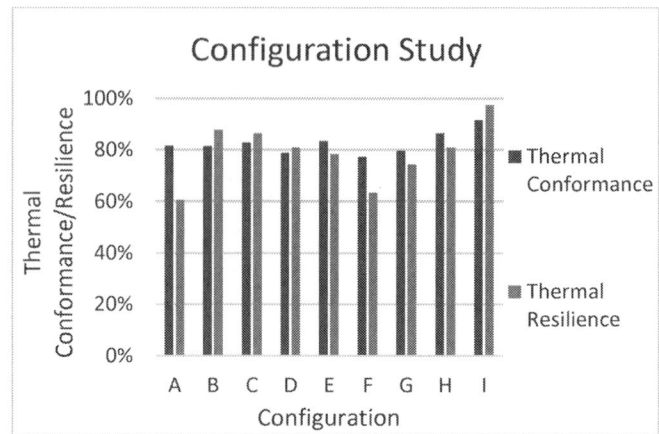

Figure 9. Results of thermal conformance/resilience for configuration study

3.3. Results of Containment

The Containment strategies study results showed significant differences when comparing them to no containment and each other. The results of the study are shown in Figure 10 with a description of the simulation in Table 2. Simulations U through W describe the most optimal configuration while simulations X through Z describes the lesser optimal configuration. Full aisle hot containment for configuration D increase thermal conformance by more than 10% compared to no containment and nearly 20% when compared to partial containment. Cold aisle containment increased thermal conformance further in each category but dropped when thermal resilience was compared. Full aisle hot containment had nearly 14% better thermal resilience for configuration K when compared to cold aisle containment. The reason for this is when a CRAH failure occurs, then the portion of the data center serviced by that CRAH does not receive enough air putting the IT equipment in that area in danger of failing.

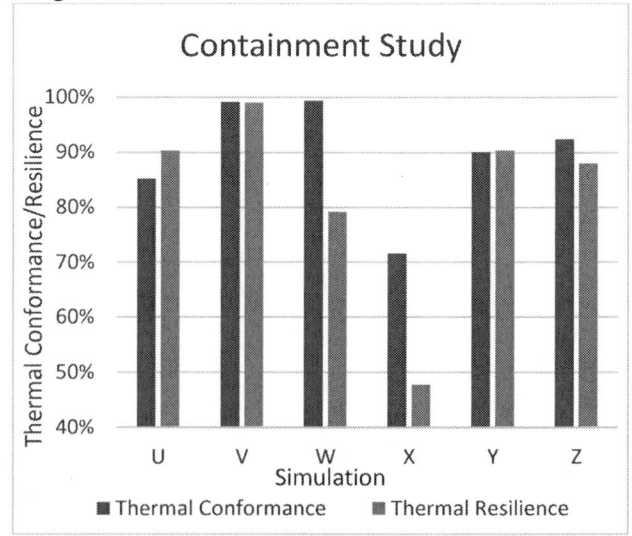

Figure 10. Results of thermal conformance/resilience for setpoint study

U	Hot - Partial - IT Switches
V	Hot -Full - IT Switches
W	Cold - IT switches
X	Hot - Partial - Power and Distributed
Y	Hot -Full - Power and Distributed
Z	Cold - Power and Distributed

Table 2. Description of containment simulations

4. Conclusions

This study used a simulation-based approach to investigate the effects of several data center parameters on a new set of metrics within the performance indicator proposed by The Green Grid. The data center parameters included CRAH setpoint, room configuration changes, and containment strategies while the performance indicator is made up of PUEr, thermal conformance, and thermal resilience. The results of the study indicated that configuration and aisle containment within a data center dramatically influences the thermal conformance and the thermal resilience with minimal effect on the PUEr. Conversely, the CRAH setpoint mainly influences the PUEr.

Acknowledgments

Fruitful discussions with Future Facilities application engineers are greatly appreciated.

References

[1] A. Shehabi, S. J. Smith and D. A. Sartor, "United States Data Center Energy Usage Report," 2016.

[2] D. Reddy, B. Setz, S. Rao, G. R. Gangadharan and M. Aiello, "Metrics for Sustainable Data Centers," *IEEE Transactions on Sustainable Computing,* vol. 2, no. 3, pp. 290-303, 2017.

[3] M. Seymour, M. Bana, D. Wang and D. Cummins, "WP#68 - THE PERFORMANCE INDICATOR ASSESSING AND VISUALIZING DATA CENTER COOLING PERFORMANCE," The Green Grid, 2016.

[4] S.-W. Ham, J. Park and J.-W. Jeong, "Optimum supply air temperature ranges of various air-side economizers in a modular data center," *Applied Thermal Engineering,* vol. 77, pp. 163-179, 2015.

[5] S. Nada, M. Said and M. Rady, "CFD investigations of data centers' thermal performance for different configurations of CRACs units and aisles seperation,"

Alexandria Engineering Journal, vol. 55, pp. 959-971, 2016.

[6] J. Niemann, K. Brown and V. Avelar, "Impant of Hot and Cold Aisle Containment on Data Center Temperature and Efficiency".

[7] Future Facilities, "6SigmaRoom," Future Facilities, [Online]. Available: https://www.futurefacilities.com/products/6sigmaroom/. [Accessed 8 October 2018].

[8] E. Samadiani, H. Amur, B. Krishnan, Y. Joshi and K. Schwan, "Coordinated Optimization of Cooling and IT Power in Data Centers," *Journal of Electronic Packaging,* vol. 132, 2010.

[9] L. Wang and S. Khan, "Review of performance metrics for green data centers: a taxonomy study," *The Journal of Supercomputing,* vol. 63, no. 3, pp. 639-656, 2013.

Transient Analysis Overshoot in Temperature for High Power Thermal Solutions

Javier Avalos
Intel Corporation
Av. del bosque 1001
Zapopan, Mexico
Javier.avalos.garcia@intel.com

Enrique Barreto
Intel Corporation
Av. del bosque 1001
Zapopan, Mexico
Enrique1.a.barreto.martinez@intel.com

Abstract

There is a trend to increase processor performance with high thermal design power consumption. Transient temperature changes from low to high power induces overshoot in temperature, generating thermal events; however, thermal solution is capable of cooling in steady state condition. This effect asserts throttling mechanism during transient state while showing temperature and power spikes. Identifying the magnitude of the overshoot in the transition from one state to another, and modifying the existing control algorithm to minimize offsets in temperature target can be key to avoid exceeding the critical temperature. The proposal is that this overshoot would be identified and remediated at initial phases of system development to mitigate any potential complaint from customers.

This work is focused on the transient behavior situation that induces overshoot in temperature, identify cases of unavoidable over temperature for a given cooling solution and fan speed control considerations to minimize this adverse condition. We present a new approach to identify the initial and final states to understand the residual energy the heatsink needs to dissipate. The hypothesis presented illustrates the theory behind this phenomena to reach the final isothermal temperature distribution when high power and high airflow are required.

Keywords

Transient, heat sink, fan speed control

Nomenclature

m, m_{die}, m_{hs}: lumped, processor, heatsink object mass (kg)
P: chipset power consumption (W)
Cp: lumped heatsink heat capacity (KJ/Kg·K)
Cp_{die}, Cp_{hs}: processor, heatsink heat capacity (KJ/Kg·K)
T, Tj, Ts: Lumped, junction, sink temperature (K)
A: surface area (m^2)
C: electrical capacitance (F)
R: electrical resistance (Ω)
V: electrical tension (V)
q: electric charge (C)
i: electrical current (A)
c: mechanical damping coefficient (N·s/m)
k: spring constant (N/m)
F: Applied force (N)
Q_{latent}: Latent heat (J)
q_{cond}: Conductivity heat transfer (W)
q_{conv}: Convection heat transfer (W)
h: heat transfer coefficient (W/m^2·K)

1. Introduction

As power consumption has increased in servers, there has been an increasing demand for thermal solution with greater heatsink sizes (Fig.1). During high power thermal system server validation it has been observed that proper heatsink cooling solutions are adequate and passing testing temperature targets for all steady state conditions. However, systems were reporting thermal events and anomalies at the autonomous stress cycle validation, suggesting insufficiency in the cooling capability, even though the fan speed control response was set to an aggressive sensitivity. To investigate the issue, the temperature was monitored under transient conditions, this was from a low power state and low fan condition for processor Thermal Design Power (TDP) and fans to full fan speed; it was realized temperature increased above its target limit for a long period of time and afterwards dropped when steady state was reached. Simulations showed that the effect was due to latent energy stored in the heatsink that had to be adjusted from low speed-low power to high speed-high power state.

Figure 1: processor heatsink examples for server sector.

Transient overshoot in temperature for electronic packages, particularly computer processors, during stress cycles is expected to occur and typically cooling thermal solution guard band is generally accommodated. However, as industry move to higher TDP values (200W up to 500W for air cooling), the overshoot effect gets magnified because of the trend in continuous increase in power consumption. Thermal designs are pushed to the limit of the steady state thermal capability, making transient response overshoot in temperature a more complex problem.

Since overshoot in temperature happens for a relative short period of time compared to operational life of the electronic package, for processors, Bios software requires programing of the length of time for whether an over the limit temperature gets reported or ignored into the system syslog event. Hence the accurate characterization of the transient time constant becomes critical for high TDP scenarios.

Robust control methodology approaches have been developed into server platforms, [1, 2]. However, the overshoot resides in the fan power saving operation point.

Fan speed methodologies are becoming more sophisticated in order to minimize the overshoot of temperatures during transients from low to high states of power consumption.

2. Single degree of freedom model (SDOF)

It is possible to understand the basic parameters governing the behavior between the transient of two different steady state conditions. For this analysis it is assumed a lumped mass object, which is considered with an internal heat source and heat being dissipated through forced convection parametrized by the heat transfer coefficient h as depicted in Fig.2. The object will contain a mass m and such material a heat capacity.

$$m \cdot Cp \cdot \frac{dT}{dt} + h \cdot A \cdot (T - T_\infty) = P \qquad (1)$$

Solution of the single degree of freedom with a constant heat transfer coefficient can be expressed as:

$$T = T_o + (T_1 - T_o) \cdot \exp(-\frac{t}{\tau}) \qquad (2)$$

Where temperatures T_o and T_1 are constants obtained at initial and final steady states temperatures, and time constant τ is which is a variable dependent on mass, heat capacity, heat convective thermal resistance; τ provides an important reference of time for how long transient response last during a transient heat power excitation.

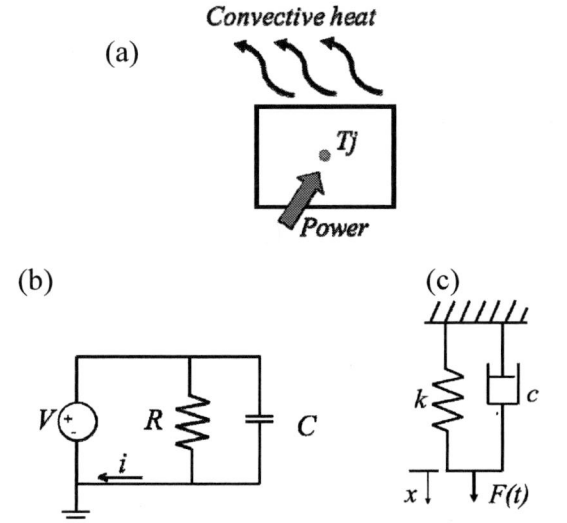

Figure 2: a) Lumped thermal model, b) electrical model analogy, c) mechanical model analogy.

Notice that the closed form solution of the transient response (Eq. 2) under constant heat transfer coefficient. This solution helps understand the time it takes from one steady state condition to another with a different power consumption, in other words, the length of the transient event. For example, for a given time τ the temperature increase/decrease will be at ~63% of the final steady state condition, at five times τ it will have reached the 99% of final state between T_o and T_1.(Fig 3)

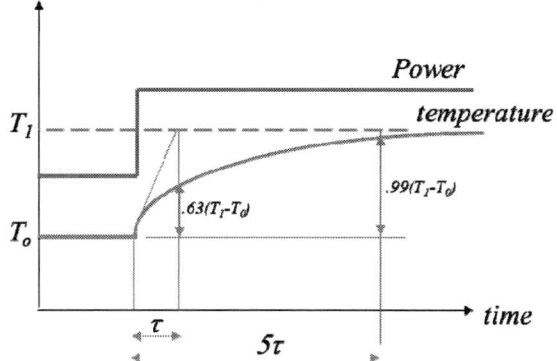

Figure 3: Single degree of freedom response to Heaviside step function excitation.

Analogous equivalent systems (Eq.3 and Eq.4) provide a better interpretation on the transient problem relating to other fields of engineering. Most importantly, the equivalent system brings a convenient alternative for testing in an easier environment, easier manipulation and measurement; particularly for the electrical models where setup building, generating excitations as well as measuring with an oscilloscope is very accessible.

Electrical analogous circuit model

$$R \cdot \dot{q} + \frac{1}{C} \cdot q = V(t) \qquad (3)$$

Mechanical analogous model

$$c \cdot \dot{x} + k \cdot x = F(t) \qquad (4)$$

3. Two time-constant model.

A more accurate model can be obtained by introducing the two time constant model or referred in this document as a two degree of freedom model (2DOF) shown in Fig. 4. This model decomposes the heat capacity of the electronic package component and the heatsink, the heat transfer between the two bodies connected with constant thermal resistance. The two differential governing equations (Eq. 6 and Eq. 7), are obtained by energy conservation at each of the components. Equation 6 corresponds to the electronic package that produces the heat to be dissipated and heat is removed by conduction to the heatsink. Ecuation 7 comes from energy balance at the heatsink component that receives the heat through conduction from the electronic package and dissipates through forced convection.

$$P - q_{cond} = m_{die} \cdot Cp_{die} \cdot \frac{dT_j}{dt} \qquad (6)$$

$$q_{cond} - q_{conv} = m_{HS} \cdot Cp_{HS} \cdot \frac{dT_s}{dt} \qquad (7)$$

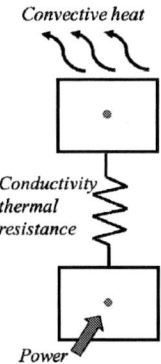

Figure 4: Two degree of freedom thermal model.

Under a constant heat transfer coefficient, the solution of the differential equations is a summation of two exponential functions, -as opposed to one as in the single degree of freedom model-, with two time constants τ_1 and τ_2. This is an improvement in accuracy compared to the previous model but is not the main reason to justify the second degree of freedom model. The two degree of freedom model is necessary to model the transient overshoot in temperature as it exhibits an inflexion of the increase in temperature followed by decrease in temperature, this effect is captured only by the two degree of freedom model.

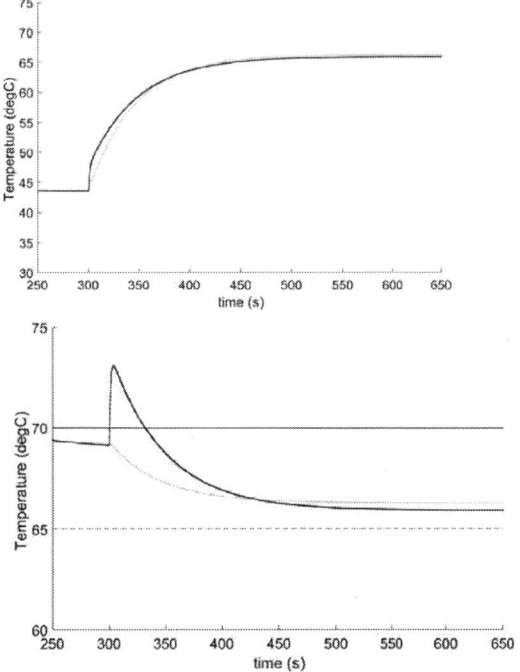

Figure 5: Temperature response of the 2DOF (blue) and SDOF (green) to Heaviside power excitation a) constant heat transfer coefficient b) step function of heat transfer coefficient.

Both SDOF model and the 2DOF model can predict the transient response in temperature behavior from one steady state condition to another. However, the 2DOF model brings a more accurate prediction at the event right after power rate changes (Fig. 5a). This is not only a marginal benefit in accuracy after that event, but a more realistic capture of the physical behavior of heatsink thermal solutions. We can see an overshoot in temperature on the transient condition from a low power consumption to a high power condition, resulting in a sudden increase in temperature followed by an inflection with a decrease until final steady state temperature condition is reached. This overshoot scenario would have been totally missed using a single time constant model, i.e., single degree of freedom model (Fig. 5b).

For scenarios where the heat transfer coefficient is not constant as a function of time, as it is the case when forced convection with fans are adjusting the rotational speed, we present a Matlab Simulink model of the 2DOF model (Fig.6). This model is used for fan speed control simulations, each constant of the model is obtained through testing characterization.

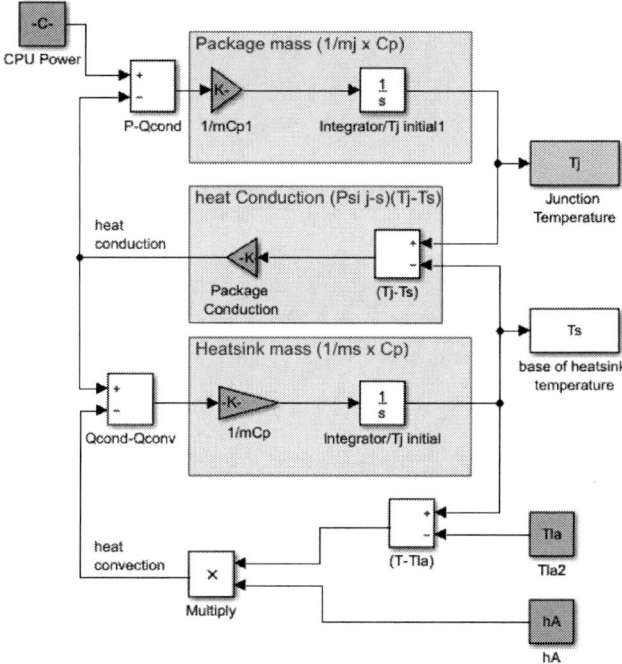

Figure 6: Simulink two time constant model.

4. Simulation and Testing

A simulation of a heatsink solution for a thermal solution designed to dissipate 250W is performed from one steady state condition into another, (Fig. 7). This is done to better explain the physics for the overshoot in temperature during transient states between low power to high power consumption.

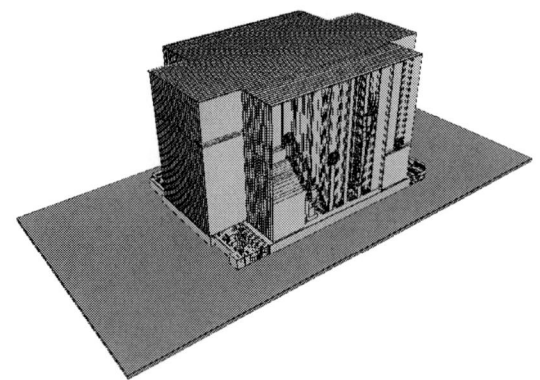

Figure 7: Heatsink thermal model (Flotherm model).

For the simulation analysis, the heatsink is meant to maintain a processor case temperature of 80°C with air environment of 35°C. The heatsink is dissipating a relatively low power consumption of 100W with the heatsink average temperature of 64°C (Fig. 8a). If we compare against the scenario with high power consumption of 300W (Fig. 8b), the fan speed requires much higher airflow creating a much higher isothermal gradient of the heatsink temperature distribution, resulting in an average temperature of 51°C. Given the mass of the heatsink of 480g and average specific heat of 900 J/Kg °C, this implies there is 5.6 KJ of latent heat that needs to be dissipated after the thermal state transition.

To demonstrate the effect in a real system, a server spread-core system with 300W processor and 24 DIMMs and 8x HDD bay area as shown in Fig. 9, was used for this test at laboratory ambient temperature of 22°C. The transient test is performed using a power virus software taking the processor from 180W power consumption up to 300W, fans ramping from 40% to 90% with full fan speed in the transition as it is seen in the Fig. 10a.

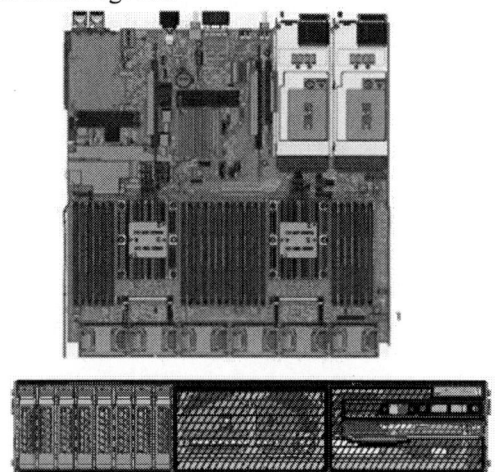

Figure 9: Testing server system board configuration.

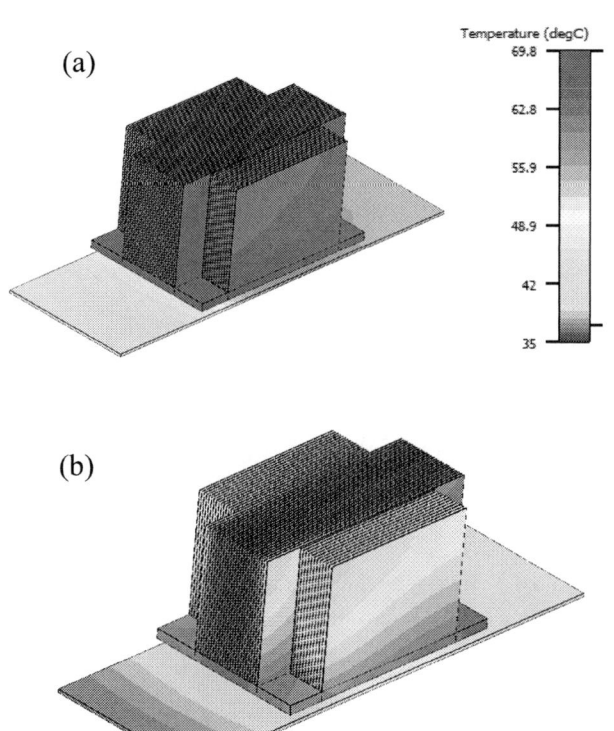

Figure 8: Heatsink flotherm simulation results (a) low power-low airflow (b) high power-high airflow speed.

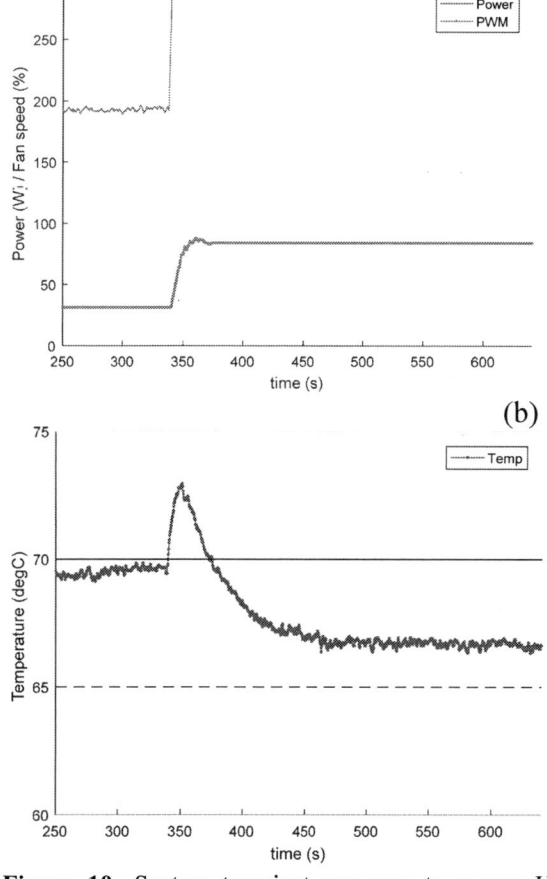

Figure 10: System transient response to power Heaviside excitation

99

During the test, the undesired overshoot in temperature is demonstrated during a transient condition from low to high power state of approximately 3°C for nearly 40 seconds as seen in Fig. 10b. This overshoot temperature is due to the sudden spike in processor bandwidth, hence, power consumption from one state into another. Specifically, when such jump in power lands at processor TDP, which is the condition where thermal solution has less margin to deal with residual heat.

From fan speed control prospective, processors are meant to adjust fan speed to maintain control temperature, such that for a low processor power consumption, low fan speed is needed for power savings. However, a fixed control temperature as opposed of allowing an operational temperature range only exacerbates the problem as seen in Fig. 11.

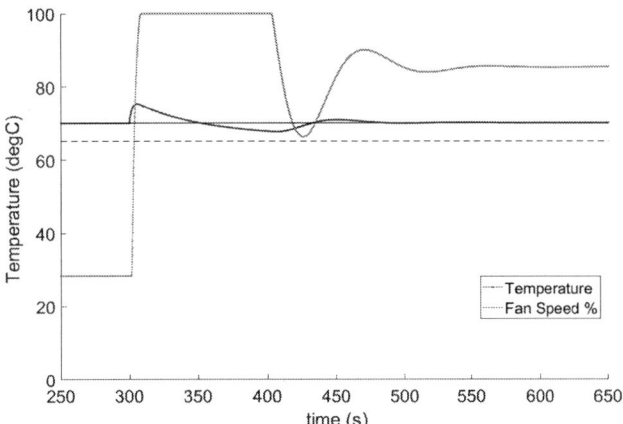

Figure 11: System transient response to power Heaviside excitation with fixed temperature control.

5. Fan Speed Control Considerations

To minimize the overshoot effect in addition to Bios settings, one must filter scenarios in temperature overshoot events during transient conditions. Fan speed control strategies can be implemented. The most obvious consisting in maintaining an aggressive fan response to increase fan speed, but careful enough to avoid oscillatory behavior given the system tracks multiple temperature sensors on board and the influence over the fan speed.

Another important consideration is the use of a tolerance for control temperature as opposed to a fixed temperature target in order to create "room" for a peak of temperature during the extraction of the heatsink latent heat. As shown in Fig. 12, a strategy to target the low control temperature target for low state of power consumption can avoid the overshoot of temperature with respect to the maximum allowable control temperature target. This "overcooling" for low state of power consumption occurs only for when fan operation is low, meaning that there is marginal extra fan power consumption in conditions where fans do not consume high power and no penalty for high state of power consumption, i.e., when

processors operate at near or at TDP. Using a temperature tolerance control allowed overcoming the overshoot and adjusting fan speed in a much shorter time, demonstrating this to be more effective than the fixed temperature control.

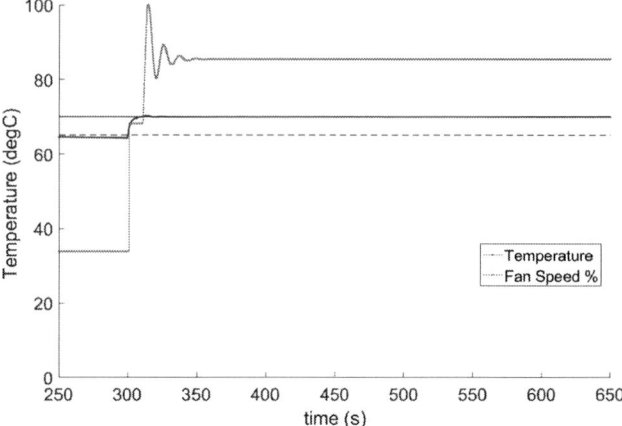

Figure 12: System transient response to power Heaviside excitation with fixed temperature control.

6. Conclusions

It is demonstrated that the two time constant model proves a more effective approach for simulating transient events. This model enables a better prediction of how fast a steady state condition can be reached, while considering the relevant variables after a transient response. Furthermore, as opposed to the SDOF model, the two time constant model can be used to predict the response of the overshoot, it is then the recommended model for fan speed control simulation.

Transient thermal cooling solution considerations were analyzed to minimize temperature overshoot in transitions. A proper thermal design with capability for cooling package at steady state is not sufficient to guarantee proper behavior during transitions. It is then necessary to maintain adequate control temperature limits for processor in addition to adequate control tuning parameters to ramp fans aggressively when bandwidth (and power) spike.

Fan capability guard band can be utilized to provide means for fan speed control to react aggressively to minimize overshoot of temperature in transitions but may be insufficient. Transient analysis helps determine the appropriate level of fan offset for overcooling at lower power levels to avoid spike in temperature and avoid exceeding temporal maximum temperature values.

References

1. Cartagena, D.G., "Applications of Control Theoretic Techinques to Improve Binsplit at the Class Test Module", Intel Assembly & Test Technology Journal, vol. 13, 2010, pp 595-605.

2. Zheng Q., Ping Z., Soares S., Hu Y., Gao Z., "An optimized active disturbance rejection approach to fan control in server", Control Engineering Practice, vol. 79, 2018, pp 154-169.

3. J.P. Holman, Heat Transfer, 9th Edition (New York 2002), pp 133-157.

Airflow Management Using Active Air Dampers in Presence of a Dynamic Workload in Data Centers

Sadegh Khalili [1], Ghazal Mohsenian [1], Anuroop Desu [2], Kanad Ghose [2], Bahgat Sammakia [1]

[1] Department of Mechanical Engineering, Binghamton University-SUNY, NY, USA

[2] Department of Computer Science, Binghamton University-SUNY, NY, USA

E-mail: skhalil6@binghamton.edu

Abstract

The dynamic nature of today's data centers requires active monitoring and holistic management of all aspects of the facility, from the applications to the air conditioning. The most significant aspect of implementing a dynamic data center is the requirement to actively monitor and manage the infrastructure assets. It is vital to ensure information technology (IT) equipment has access to sufficient air (provisioned) at a proper temperature to assure their optimal and continuous operation. Hot air recirculation, elevated fan speed, and hot spots are known consequences of an under-provisioned cold aisle. On the other hand, over-provisioning a cold aisle can lead to a significant energy loss due to cooling air bypass. In addition, the number of active servers in an aisle and their workload levels may be varied by load balancers due to short or long-term IT load changes. This demonstrates the need for an active airflow management scheme that is able to respond to changes in airflow demand in different aisles of a data center. In this study, remotely controllable air dampers are implemented to regulate airflow delivery to a cold aisle containment (CAC) during workload changes in a data center. The energy saving opportunities are investigated and practical considerations are discussed.

Keywords

Airflow Management, Active Control, Air Damper, Dynamic Workload, CRAH, Blower, Chiller, PUE, MLC.

Nomenclature

CAC	Cold aisle containment
COP	Coefficient of performance
CRAC	Computer room air conditioner
CRAH	Computer room air handler
IT	Information technology
ITE	Information technology equipment
MLC	Mechanical load component
OAR	Open area ratio
p	Pressure, Pa (in wc)
PUE	Power usage effectiveness
Q	Flow rate, m^3/s (cfm)
SLB	Smart Load Balancer
t	Time, minute

1. Introduction

Data centers play a vital role in many aspects of modern life such as banking, entertainment, business, communications, government, security, healthcare, airlines, education, etc. These mission-critical facilities are responsible for storing, processing, and managing huge amounts of data, and have been growing fast to propel the expanding demand for online services. Consequently, the share of worldwide energy consumption for data centers has grown significantly over the past decades. One of the most common concerns in the data center industry is the cooling efficiency. Currently, air-cooling is the most popular technology in data centers due to its proven high reliability, compatibility with the environment and lower initial and maintenance costs. In a large portion of air-cooled data centers, the space underneath a raised floor is pressurized by cold air from computer room air handlers (CRAH) or computer room air conditioners (CRAC) and the supplied cold air is delivered to server racks via perforated floor tiles. In such data centers, cold air can be delivered to where it is needed by simply placing a perforated tile at that location. The fans inside the information technology equipment (ITE) are responsible for drawing cold air from the adjacent cold aisles to cool their internal components. The airflow demand of ITE is a function of their fan speed and depends on various parameters such as vendor, model, configuration, environmental conditions (e.g. temperature and differential pressure across the ITE) and the workload.

In recent years, implementation of containment systems (either hot or cold aisle containment) has become an important energy-saving strategy to minimize the mixing of hot and cold air. By minimizing the mixing of cold and hot air streams, the available cooling capacity and data center's thermodynamic efficiency increase. Impacts of hot and cold aisle containment on data center efficiency and power usage effectiveness (PUE) are discussed extensively in the literature [1–9]. Minimizing the mix of cold and hot air streams allows for a higher temperature setpoint of cooling units which leads to significant savings in chiller plant energy. In addition, elimination of cold air bypass allows for further savings by lowering blower speeds in the cooling units. Based on affinity laws, blower power is proportional to the cube of airflow rate. Therefore, optimizing the airflow rate can significantly decrease fan energy consumption. Ideally, the supplied airflow rate to an aisle should match the ITE airflow demand precisely. However, from a practical standpoint, a perfect airflow match may not guarantee the elimination of hot air recirculation. The distribution of supplied air across the containment system is often not uniform [8,10,11]. Although the containment system self-balances to a large extent, gaps in cabinets and the containment system along with local negative/positive pressures in the containment system allow for local recirculation. If the supplied airflow rate is too close to the ITE demand in a CAC with significant leakage paths, a portion of cold air would bypass the ITE and leak to the hot aisle (desired leaks), but there could also be hot air leakage into the containment system [7,12]. For instance, Khalili et al. [11] showed that the racks farther from the cooling unit in a CAC system may experience recirculation through a gap under the cabinets while bypass of air was observed through the gap below the racks that are closer to the cooling unit. Therefore, slight over-provisioning is recommended in addition to proper containment sealing. The amount of recommended excess air is specific to the installed ITE and the quality of sealing in the containment system.

Heterogeneous data centers have become popular as they offer efficient use of available resources, increased reliability, higher security, and lower cost. These data centers utilize a wide range of networking, servers, and storage equipment purchased from various vendors. Hence, the airflow requirement of the equipment and its response to environmental conditions and workload can be considerably different. Also, the distribution of the IT equipment between aisles can be significantly different in these data centers. In addition, most data center workload demands exhibit daily or weekly periodic patterns [13–18] in which the demand for services can vary significantly. Furthermore, upgrading ITE may alter the airflow demand in the corresponding aisle. Therefore, airflow demand in different aisles can vary significantly and be a function of time. Heterogeneous data centers usually utilize a common open supply airflow path (e.g. raised floor) to provide flexibility to various clients. A disadvantage of traditional raised floors is that cooling units have no control over the destination of the supplied cold air. As a result, some of the aisles with a low airflow demand may receive excess airflow (i.e. are overprovisioned). There are limited solutions to balancing the airflow rate between aisles such as implementing perforated tiles with different open area ratio (OAR) or adding physical obstructions to the raised floor. Clearly, the above solutions require significant physical work and downtime. Adding to this dilemma is the fact that, these solutions are quasi-permanent and cannot address variation in airflow demand due to a transient workload. Therefore, an active approach for managing and balancing airflow is not only a way for operating such data centers efficiently but eases and reduces the cost of upgrading a data center by eliminating the need for replacing perforated floor tiles.

One approach to managing airflow locally is the deployment of active tiles. An active tile utilizes multiple fans mounted on the anterior side to boost the airflow by increasing air draw from the plenum. Arghode et al. [19] studied the benefits of using active fan tiles for increasing the efficiency in open and enclosed cold aisles. They concluded that implementation of active tiles can improve the uniformity of temperature in open and contained cold aisles compared to an under-provisioned cold aisle with passive tiles. However, more hot air entrainment was observed in the presence of active tiles. This higher entrainment can be due to jets of air with high velocities resulting from the fans in active tiles. Active tiles can also be utilized locally to increase the flow rate of tiles that are affected by an underfloor vortex. High-velocity jets of air introduced by air handler units (AHU) or the presence of underfloor obstructions can introduce vortices and stagnant regions in a plenum [20,21]. Aside from all the benefits, installing active tiles may affect the flow rate of the adjacent passive tiles due to the suction created by the active tile fans. In addition, the fans in active tiles are relatively small and consume power which increases the operating costs in data centers. Also, a large portion of the tile surface is blocked for decreasing fans recirculation and installing large batteries to

provide power during outages. This blockage increases the resistance on the flow through the tiles when fans are not operated. An alternative approach for local airflow management is deploying air dampers. Remotely controllable air dampers are installed below directional tiles and allow adjusting the flow rate by controlling the angle of the dampers' vanes (see Fig. 1). The directional tile directs airflow toward the face of the adjacent racks and increases the rack capture index. In normal operation mode, the required power for operating the damper is supplied by a power over ethernet (POE) switch via a network cable which eliminates the need for separate power cabling. In addition, the electric motors in these dampers only operate to adjust the angle of the vanes when needed. This increases the motors' lifetime and allows for smaller backup batteries. Also, the batteries can last longer in case of a power outage. However, unlike active tiles, the flow rate of individual tiles cannot be boosted by using active dampers.

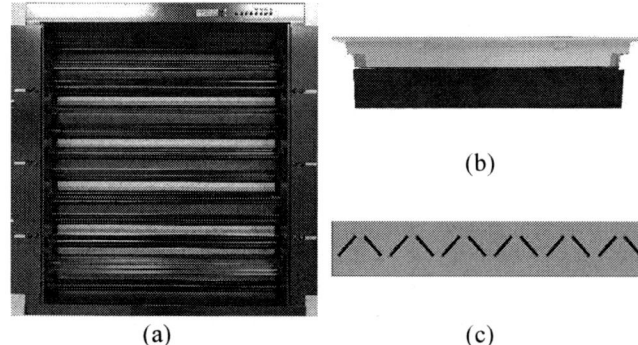

(a)   (c)

Fig. 1: Air damper: a) top view of the damper, b) side view of the damper mounted under a perforated floor tile, c) schematic of the damper's vanes.

In this study, the impacts of employing remotely controllable dampers on energy efficiency in the presence of a dynamic load are investigated at the aisle level. Resultant energy savings in blower and chiller powers are calculated. Also investigated are various approaches for turning off non-utilized servers, namely turning off all the servers in preselected racks, and turning off multiple servers at the bottom of each rack. In addition, thermal imaging and a smoke test are utilized to detect potential locations prone to recirculation.

2. Modeling of IT Load Variation in Data Centers

Dynamic server provisioning can be employed to lower power consumption and airflow demand, and increase system utilization. The basic idea is to consolidate the workload from several servers with low utilizations onto fewer servers. This allows saving energy by turning off or enabling sleep state in the servers that are not needed for satisfying the workload requirements. Additional servers can be turned back on when the demand increases [15,16]. In the case of virtualized environments, the unneeded virtual machines can simply be released back to the cloud to save on rental costs. However, the sleep and off states can affect the response time in data centers without a sophisticated workload management system [15,16]. The CPU utilization over a two weeks period for two random servers in a commercial data center is presented in Fig. 2. A quick visual inspection reveals that CPU utilization remains very low over long periods. Also, a periodic behavior is observed which allows for load prediction. Therefore, load balancing can be utilized to for provisioning ITE efficiently.

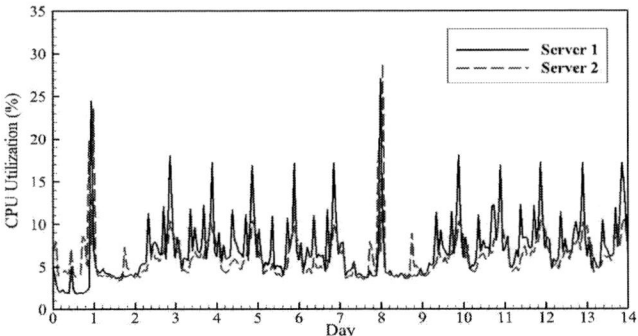

Fig. 2: CPU utilization data over a two weeks period in a commercial data center.

Stachecki and Ghose [15] proposed a Smart Load Balancer (SLB) - a fully automated solution - for provisioning IT capacity to match the instantaneous offered workload that results in no performance loss against a baseline design where all servers remain active but operate at a lower CPU utilization. Figure 3 represents the variation of actual utilization trace of a data center operator from low IT load hours to the peak load period. In this trace, the utilization of the system stays below 20% for two-thirds of the duration of the trace. In order to consolidate IT load onto a fewer number of serves, the SLB can be used to schedule jobs. There are two variants of SLB depending on the configuration parameters, namely SLB-LT and SLB. In SLB-LT lower utilization thresholds (40% - 60%) are used to reduce the impact on the tail latency. During the dormant period of the trace, SLB-LT schedules the incoming workload onto 41% of the initial working set of servers and the remaining 59% of servers are powered down. The second variant, SLB, uses higher CPU utilization thresholds (50% - 60%) than SLB-LT and schedules the incoming workload to a fewer number of servers compared to SLB-LT. With the second variant, the results showed that 66% of the servers were not utilized and therefore, could be turned off during the dormant period of the trace.

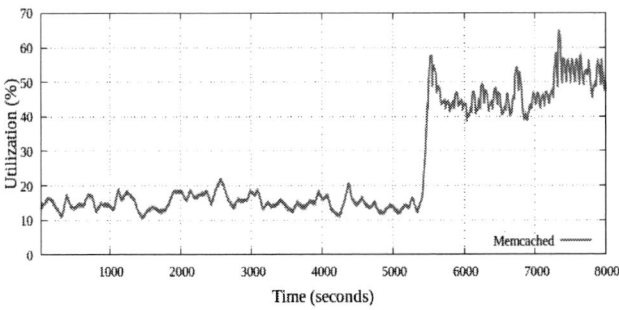

Fig. 3: Experimental CPU utilization as a function of time for one of the tested servers based on a real data center trace.

A challenge in dealing with the dynamic workload is the possibility of abrupt changes in load. Several instances of load spikes have been documented during important events, such as elections, sporting events, natural disasters, Black Friday shopping, slashdot effects, etc. To handle sudden spikes as well

as short-term fluctuations in the load, SLB keeps a fixed number of servers powered on but not utilized. In the above experimental runs, 14% of the servers had been kept online to mimic avoiding a potential lag due to the setup time, and violation of the Service Level Agreement (SLA). In this study, a more conservative approach is chosen by turning on/off 33% and 45% (instead of 59% and 66% in SLB-LT and SLB, respectively) of the servers. This increases the available capacity further for handling large unpredicted spikes in the load and minimizes risk of SLA violation. It should be noted that higher energy savings can be achieved by utilizing SLB or SLB-LT in which a larger portion of servers can be powered off compared to the studied cases in this paper.

3. Experimental Setup

The experiments are conducted in the ES2-Data Center Laboratory at Binghamton University. The lab is a 215 m^2 (2,315 ft^2) space with a 0.91 m (3 ft) deep raised floor, and is equipped with two chilled water-based CRAH units, which are rated at 114 kW (32 tons) of cooling capacity. The nominal maximum airflow rates of the cooling units are 16,500 cfm and 17,500 cfm for CRAH 1 and CRAH 2, respectively. The CRAHs are equipped with variable frequency drives so that supplied airflow rates can be modulated. The layout of the lab is shown in Fig. 4. The IT equipment rows are set up in an alternating hot aisle/cold aisle arrangement. The cold aisles in aisles C and D are contained. In this study, air dampers are installed bellow directional tiles in aisle C. Sixteen racks containing 272 2-RU servers from different vendors and generations are deployed in aisle C. Table 1 shows the list of installed ITE in this aisle and their corresponding airflow demand for an inlet air temperature of 20 °C. The airflow demand of each model was measured by mounting a server of the corresponding model on a flow bench designed in accordance with AMCA 210-99/ASHRAE 51-1999. The experimental details are similar to the characterization process described in [22]. It should be mentioned that the number of installed servers in the racks varies from 14 to 20 servers per rack (equipment density at the beginning of the aisle is higher).

Table 1: List of ITE and their airflow demand in aisle C.

IT Equipment	Required Airflow Rate (cfm)	Total Number	Total Airflow Demand (cfm)
Dell PE R530 - Conf.1	35	79	2765
Dell PE R530 - Conf.2	37	32	1184
Dell PE R520	33	61	2013
Dell PE R730	33	6	198
HP PL DL380/385 G6	34	64	2176
HP PL DL380/385 G7	35	13	455
Dell PE 2950	44	3	132
Dell PE C2100	34	14	476
Network Switch	29	16	464
Sum:	**286**		**9620**

The tiles in aisles B and D are covered in this study while some of the tiles in aisles A and E were left open to provide air for some of the essential networking equipment for operating the data center lab. Three Bapi ZPT-LR pressure sensors with a measurement range of 1″ wc (248.84 Pa) and accuracy of ±0.25% of the range are installed at the top of the racks C1-1, C1-4 and C1-8 (see aisle C in Fig. 4) for monitoring pressure inside the CAC. A flow hood (CFM-850L) is used to measure airflow rate of the tiles. The accuracy of flow hood measurements is analyzed by mounting it on the flow bench and comparing its data with flow rates of the flow bench at various flow rates. It is found that the flow hood shows between 7% to 12% higher flow rates compared to the flow bench. To minimize the error in airflow measurements, flow hood readings are corrected based on flow bench measurements. The maximum discrepancy between the flow bench and flow hood data was reduced to 2% via the correction. This allows a more accurate assessment of provisioning level based on the data presented in Table 1 that was measured using the flow bench.

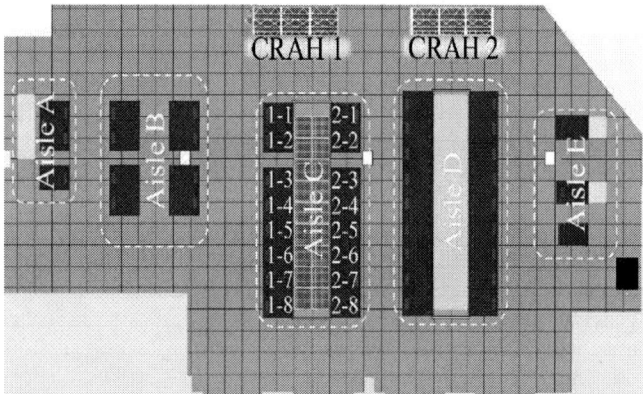

Fig. 4: Layout of the ES2 data center lab at Binghamton University.

A Fuzzy feedback controller is designed to adjust the OAR of the dampers based on pressure measurements in aisle C. Generally, fuzzy controllers are built from a set of if-then rules. An important advantage of fuzzy controllers is that a mathematical model is not necessary [23]. Hence, a fuzzy controller is an ideal choice for controlling complex systems in which a mathematical model of the system is not available or is computationally prohibitive. A CFD simulation for a data center is computationally expensive and a real-time simulation is not feasible with current resources. Also, characterization of leaks through gaps in the cabinets, containment, and raised floor is cumbersome, therefore, a fuzzy feedback controller is well-suited for the application in this paper. The designed controller adjusts the openness of the dampers via Simple Network Management Protocol (SNMP) with various rates based on an error signal. The error signal is defined as the difference between measured pressures and an ideal pressure range. For small error signals, the controller changes the openness of the damper with fixed small steps to avoid overshoot in CAC pressure. For larger error signals, the controller measures the response of the pressure in the CAC by making an arbitrary change in the OAR of the dampers first. Subsequently, the controller estimates the next OAR by a linear extrapolation based on the latest response of CAC pressure to the OAR. In this study, the goal is to reach and maintain a neutral provisioning state while 1 Pa over-provisioning of the CAC is allowed ($0 \ \text{Pa} \leq p_{ideal} \leq 1 \ \text{Pa}$).

4. Test Procedures

The developed fuzzy control system is tested in increasing and decreasing IT load cases in which some of the servers were powered off and powered on, respectively. Based on an analysis described in section 2, implementing SLB-LT and SLB allows turning off 59% and 65% of the servers during the dormant period of the trace. As mentioned before, a more conservative approach is considered in this paper in which 33% and 45% (instead of 59% and 65%) of the servers are switched on/off. In the first scenario, the impact of under-provisioning in aisle C is visited briefly. Smoke tests and thermal imaging is used for detecting potential areas prone to recirculation. Consequences of under-provisioning an aisle are explored in numerous studies, so is not discussed in this paper extensively. Interested readers may refer to [10,11,24–27].

In scenario 2, 6 racks (C1-6 to C1-8 and C2-6 to C2-8) are turned on/off to mimic variation of IT airflow demand due to sudden changes in IT load. The response of control system and pressures in the CAC are monitored from the initial steady state (when 33% of the servers were powered off) to the second steady state after powering on all the servers. Next, a decrease in IT load is mimicked by powering off 33% of the servers and the transient response of the system is monitored. A similar test procedure is used in scenario 3 except that 8 racks are powered on/off to mimic the status change in 45% of all the servers. After testing the performance of the control system in scenarios 2 and 3, 8 servers in the bottom half of all the racks are turned on/off to investigate the effect of location of off servers in scenario 4. The results are compared with scenario 3. Table 2 summarizes the described test scenarios.

It is worth mentioning that a drift in the installed differential pressure sensors was observed in some of the tests. Thus, frequent calibration may be required for proper operation of the controller. In addition, it should be noted that the response time of the system depends on the response of servers' fan speed to a change in IT load or environmental conditions. However, no significant change in fan speed of active servers is observed in the studied scenarios.

Table 2: Summary of studied scenarios

Scenario	Provisioning Level	Location of off servers	% of off Servers
1	Under-provisioned	End of aisle C	22%
2	Provisioned	End of aisle C	33%
3	Provisioned	End of aisle C	45%
4	Provisioned	Bottom of the racks	45%

5. Results and Discussions

5.1. Scenario 1

The goal of this scenario is to determine potential leak paths in aisle C implementing thermal imaging via an infrared (IR) camera and smoke visualization. The IR camera shows the apparent surface temperature of the objects. However, in a steady state, it can be assumed that the surface temperatures in the thermal images represent airflow temperature passing over the surfaces when heat generation in the object is not significant. In this scenario, the aisle is slightly under-provisioned to ease detection of recirculation paths. Initially, racks #5 to #8 in rows C1 and C2 (four racks in each row) were turned off and the aisle was provisioned. After assuring a steady state, racks #5 and #6 in both rows are turned on while the OAR of the dampers remains fixed. This dropped the CAC pressure to -2 Pa, -1 Pa and -1 Pa at the beginning, middle and end of the aisle, respectively. Figures 5a and 5b show thermal images captured from the end of the aisle after the initial steady state (neutrally provisioned aisle). Slightly elevated temperatures are observed in servers of rack C1-8 which is due to lack of blank fillers in the front panel of these servers. Temperatures in the rest of the aisle are cool. It should be mentioned that warmer temperatures corresponding to blanking panels are due to the contact of blanking panels with hot air in the hot aisle.

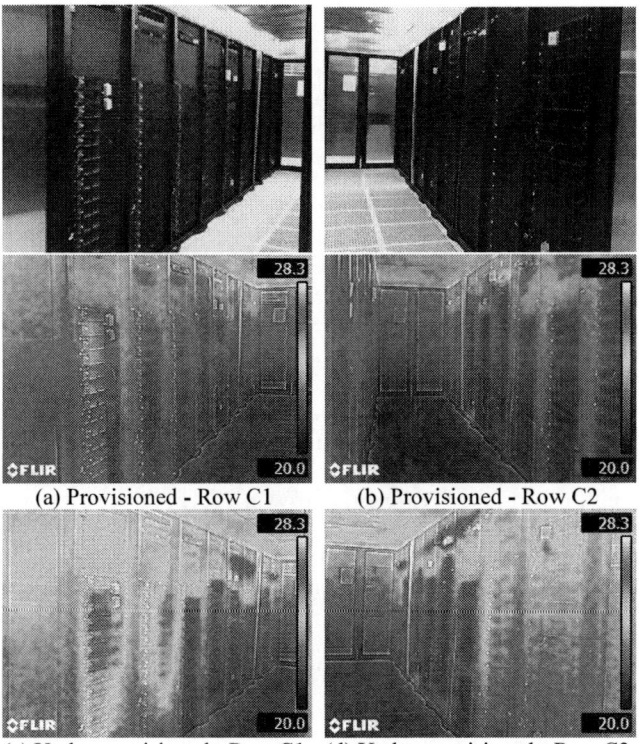

(a) Provisioned - Row C1   (b) Provisioned - Row C2

(c) Under-provisioned - Row C1   (d) Under-provisioned - Row C2
Fig. 5: Thermal images in the neutrally and under-provisioned CAC in scenario 1.

The thermal images in Figs. 5c and 5d are captured when aisle C is under-provisioned. High-temperature areas in front of racks # 7 and #8 demonstrate that the powered-off servers are clearly a significant recirculation path although the magnitude of differential pressure across the containment was small. The leaked hot air mixes with cold air supplied by the tiles and slightly elevates the overall CAC temperature. Also, it is seen that surface temperatures increase across the racks' height. The colder surface temperatures at the bottom of the racks are due to the type of the floor tiles which direct airflow towards the face of the racks. The impingement of air to the front panel of the servers causes a higher pressure there, specially for the server that are closer to the tiles. This higher pressure hinders recirculation through these servers. Figures 5c and 5d also reveal another recirculation path at the top of the racks. This is where intake ducts connect the intake of rear mounted network switches to the CAC. Further investigation

reveals that hot air leaks into the CAC through the brush grommet used between the intake ducts and the switches due to the negative pressure in the CAC as shown in Fig. 6a. The total flow rate through the tiles was measured as 5880 cfm and 5986 cfm in the first and second steady states which shows a slight increase while the OAR of the dampers was fixed. The higher tile flow rate in the second steady state is due to the increased airflow demand in the aisle after turning on the servers and the resulting negative pressure inside the CAC.

(a)   (b)   (c)
Fig. 6: Smoke visualization reveals the potential recirculation paths: a) recirculation through the intake duct of network switches, b & c) hot air sucked into neighboring servers.

To further investigate the impact of powering off the servers in a rack, another case is studied in which servers at the bottom of all the racks are turned off instead of powering off all the servers in some of the racks. Similar to the previous test in this scenario, it is observed that hot air recirculates through the servers that are powered off. In this case, smoke tests showed that the leaked hot air was sucked into the active servers above the leak path (see Fig. 6b). This concentrates the impact of hot air recirculation onto a few servers that are near the leak path and creates hotspots in a specific height of the racks. Similar behavior was observed when servers at the top of the rack were turned off as shown in Fig. 6c. Overall, the results demonstrate that even a slight under-provisioning can cause recirculation through the servers that are powered off and create hot spots inside the containment.

5.2. Scenario 2

In this scenario, the performance of the controller is tested in presence of an IT load variation. Figure 7 presents the variation of dampers' OAR and pressures at the beginning, middle and end of aisle C during the test. Between $0 < t < 10$ minutes, the servers in 6 racks (C1-6 to C1-8 and C2-6 to C2-8) were off and the pressures were within the ideal range. It should be noted that the step changes in pressures are due to the resolution of pressure measurements which is 1 Pa in this study. At $t = 10$ minutes, these racks were turned back on and the controller started to adjust dampers OAR to respond to the resultant negative pressures in the CAC. The controller response varies with respect to the magnitude of under/over-provisioning as seen in Fig. 7. For example, when the magnitude of the error signal is equal to or more than 3 Pa, the controller decreases the OAR with larger steps calculated based on the previous response of the system to smaller steps. Figure

8 demonstrates that the recirculation through servers shown in Fig. 5c is eliminated by properly provisioning the CAC. The total response time of the system is less than 8 minutes in this part. It should be mentioned that the controller monitors the pressures every 10 seconds in normal operation but sleeps 90 seconds after each change in the OAR of the dampers to allow sufficient time for the system to reach a quasi-steady state before assessing the pressure conditions again. This delay increases the stability of the control system and minimizes the chance of an overshoot due to fan speed changes during set-up time of the servers in which there is a temporary spike in servers fan speed.

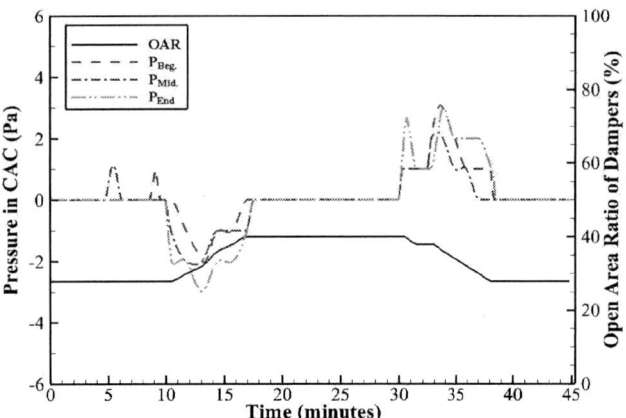

Fig. 7: Variation of pressure and OAR in scenario 2.

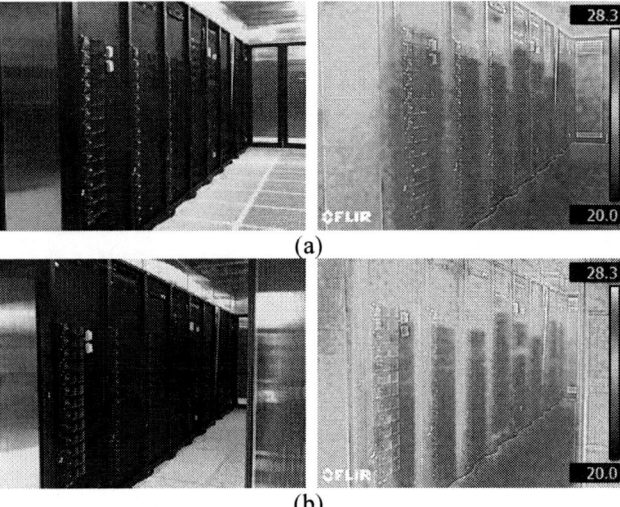

(a)

(b)

Fig. 8: Thermal image of row C1: a) before and b) after turning on the servers in scenario 2.

After reaching the second steady state at t = 30 minutes, the same 6 racks were turned off to mimic the response of a load balancer to a drop in IT load. Figure 7 shows an initial steep change in the pressure by 2 Pa at the end of the aisle (where the racks were turned off). As the pressure exits the ideal range, the controller responds by decreasing the OAR of dampers by 2%. This slight decrease along with the diffusion of excess air in the CAC brings the pressures within the ideal range temporarily. After some time, the excess air builds up in the CAC and increases the pressure which is followed by the controller response until the system reaches the final steady state of this

scenario. Similar to the increasing load case, the overall response time of the system to the IT load drop is less than 8 minutes. It should be mentioned that no significant change in the fan speed of active servers was observed in this scenario.

5.3. Scenario 3

The test procedure in this scenario is similar to scenario 2 except 8 racks (C1-5 to C1-8 and C2-5 to C2-8) were turned on/off to mimic a more aggressive load balancing due to increasing and decreasing IT loads. The transient response of the controller and pressures at the beginning, middle and end of the aisle are presented in Fig. 9. As expected, the magnitude of pressure changes and the OAR of dampers are higher than pressure variations in scenario 2, which is due to the larger number of servers that participate in handling the IT load change. In addition, the system's overall response time is increased to 10 minutes compared to 8 minutes in scenario 2. It is noted that the pressures at the end of the test are at the upper limit of the ideal pressure range due to the higher OAR of the dampers at the end of the test (23% compared to 21% at the beginning of the test). Table 3 compares the measured tile airflow delivery and estimated total airflow demand in the aisle. The estimated airflow demands are calculated based on the number of active servers in each steady state and the data presented in Table 1. The difference between the supplied flow rate through the tiles and ITE demand is presented in the "Difference" row of this table. The differences can be due to the bypass of airflow through containment gaps and the servers that are powered off, as well as uncertainty in the pressure and flow rate measurements. The overall uncertainty of the measurements does not exceed 6%. This means the primary contributor to the "Difference" row in Table 3 is the bypass of airflow through the servers and containment gaps. Also, this difference is higher in the final steady state which is due to the higher pressure in the CAC in the final steady state of this scenario. Figure 10 shows that that the aisle is properly provisioned and no recirculation is detected.

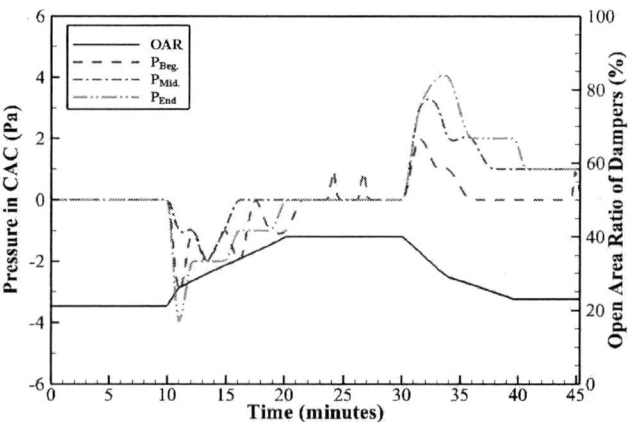

Fig. 9: Variation of pressure and OAR in scenario 3.

Table 3: Estimated and measured flow rates in scenario 3

	Initial S.S.	Second S.S.	Final S.S.
Tiles' Flow rate	5935 cfm	9927 cfm	6164 cfm
Est. airflow demand	5430 cfm	9620 cfm	5430 cfm
Difference	505 cfm	307 cfm	734 cfm
% Overprovisioning	9%	3%	13%

(a)   (b)

Fig. 10: Thermal image of row C1 after turning on the servers (the second steady state) in scenario 3.

5.4. Scenario 4

In this scenario, 8 servers at the bottom of all the racks of aisle C are turned off (45% of all the servers). The controller was able to manage airflow delivery to the CAC successfully. The results were similar to scenario 3 except that the OAR of dampers in the initial and final steady states was slightly higher. Graphs are not presented here for the sake of space. Table 4 presents the tile flow rates in the initial and the second steady states. By comparing tile flow rates between the initial steady states in tables 3 and 4, it is perceived that turning off the servers at the bottom of the racks increases air bypass through these servers. As mentioned earlier, the installed tiles direct airflow toward the face of the racks which creates a higher pressure at the inlet of the servers. This effect is stronger at lower elevations where servers are closer to the tiles, i.e. servers at the bottom of the racks. Therefore, cold air bypass through the powered-off servers in this scenario is more than the corresponding bypass in scenario 3.

Table 4: Estimated and Measured flow rates in scenario 4

	Initial S.S.	Second S.S.
Tile flow rate	6424 cfm	9950 cfm
Est. airflow demand	5495 cfm	9620 cfm
Difference	929 cfm	330 cfm
% Overprovisioning	16%	3%

6. Energy Saving Opportunities

Iyengar and Schmidt [28] presented an analytical model for thermodynamic characterization of cooling systems in data centers using experimental data and empirical equations. In a case study, they found the chilled water air handler as the second largest energy drain on the cooling system after the chiller (see Fig. 11). A similar order is reported in [12] and [29].

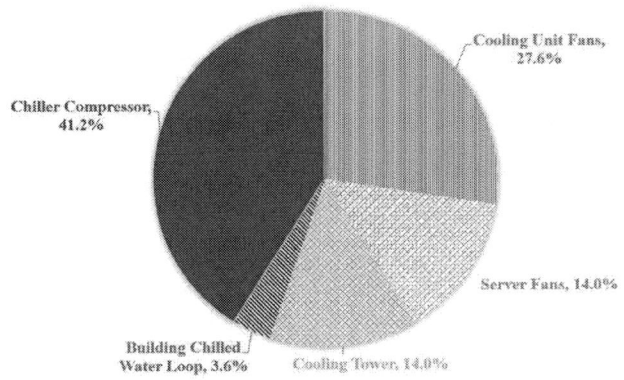

Fig. 11: Cooling energy breakdown for the case study example in [28].

Comparing tile flow rates in Table 3 shows that air delivery to aisle C can be decreased by approximately 40% via implementing the controller. The most significant energy saving associated with a decrease in airflow delivery to an aisle comes from adjusting the fan speed of cooling units. The energy required for moving air by a fan is proportional to the cube of the airflow rate.

$$\frac{P_2}{P_1} = \left(\frac{Q_2}{Q_1}\right)^3 \tag{1}$$

According to affinity laws (1), cutting airflow rate by 40% will reduce CRAH's blower power by 80% approximately. However, many data centers do not dedicate a cooling unit per aisle. The cooling units are usually oversized to serve multiple aisles and benefit from higher-efficiency larger blowers or provide cooling redundancy in data centers. In addition, a portion of the supplied air bypasses through raised floor gaps and does not reach the cold aisles. In practice, the percent of the decrease in a CRAH's flow rate is smaller than the decrease in an aisle's airflow demand. Therefore, the actual power saving is expected to be less than 80%. The actual saving can be estimated by making some assumptions. To have an accurate estimate, raised floor leaks should also be taken into account. The amount of air bypass through raised floor gaps depends mainly on the pressure in the plenum. A reasonable assumption is that this leakage remains fixed if the pressure in the plenum is controlled via adjusting blower speed of the cooling units. As the OAR of tiles in aisles A and E are fixed, it can be assumed that flow rates of these tiles do not change if the plenum pressure is fixed. Maintaining a fixed pressure in the plenum also minimizes the impact of changing the OAR of dampers in an aisle on the airflow delivery to other aisles. However, maintaining a fixed pressure in the plenum requires a separate controller that controls the fan speed of cooling units, and is out of the scope of this paper. The required CRAH's flow rate can be estimated by

$$Q_{CRAH} = Q_{Aisle\,C} + Q_{Aisle\,A} + Q_{Aisle\,E} + Q_{R.F.\,Leaks} \tag{2}$$

in which $Q_{R.F.\,Leaks}$, $Q_{Aisle\,A}$ and $Q_{Aisle\,E}$ are rate of air leak through the raised floor, air delivered to aisle A and E, respectively. The above flow rates remain constant when plenum pressure is fixed. Using (2), $Q_{R.F.\,Leaks}$ can be calculated. Following the assumptions above, required CRAH airflow rate after turning off servers is calculated, showing a 25% decrease compared to when all the servers were powered on. This decrease in CRAH's airflow translates to 58% drop in the CRAH's power according to Eq. (1).

Another significant benefit of airflow management is in the chiller plant. In the ES2 Data Center Lab, the chilled water is supplied from the building's chiller. Thus, a direct measurement of savings was not possible. Instead, a typical value for the coefficient of performance (COP) can be used to estimate the chiller power. The chiller power can be calculated via (3) [30]:

$$P_{Chiller} = \frac{Removed\ Heat}{COP_{Chiller}} \tag{3}$$

Turning the servers off and consolidating workload by the load balancer decreases the number of idling servers but increases the power of active servers. However, there is a

107

logarithmic relationship between CPU utilization and the server's power typically such that operating at a higher CPU utilization increases the efficiency of a server. The power consumption measurements in different scenarios of the study showed the overall IT power decreases by consolidating workload and turning off excess servers. The average rise in air temperature due to IT heat dissipation in aisle C can be calculated using heat balance:

$$\Delta T_{air} = \frac{\text{IT Heat Dissipation}}{\dot{m}_{air} \cdot c_p} \quad (4)$$

Assuming negligible change in air properties, equation (4) gives an estimate on the return air temperature to the CRAH unit for a fixed supply air temperature of the CRAH unit. Powering off the non-needed servers decreases the total IT power and airflow demand in the aisle. Two cases are compared in Table 5. In case 1, all the servers in aisle C are active and operate at 15% CPU utilization. Case 2 represents aisle C with active airflow management and a smart load balancer. Table 5 shows that despite the decrease in the total heat dissipated by ITE, the overall temperature rise is higher in case 2. This is due to the cut in the supplied airflow rate to the aisle and the higher workload on the active servers. A relatively high rack air temperature rise leads to a more efficient heat transfer in the CRAH's coils and higher fluid temperature returning to the chiller evaporator which improves overall cooling performance and increases the COP of the chiller. Breen et al. [31] concluded that the COP gain due to an increased rack air temperature rise is considerable when this rise is below 12.5 °C which is the case in Table 5. In this study, a COP of 3.0 is assumed for the base case and an increase of 0.2 in COP per °C increase in rack air temperature rise. Implementing the assumptions above, the chiller power can be calculated using (3). The results show 36% saving in the chiller power consumption.

Table 5: Comparison of IT power without and with implementing controller and load balancing for the trace presented in Fig. 3.

Case #	CPU Util.	# of Active Servers	IT power (kW)	Airflow Rate (cfm)	ΔT_{air} (°C)	COP_{Ch}	Chiller Power (kW)
1	15%	271	44.8	9927	8.1	3	14.9
2	65%	161	30.4	5935	9.2	3.2	9.5

The power usage effectiveness (PUE) corresponding to cases 1 and 2 in Table 5 can be calculated via (5)

$$\text{PUE} = \frac{1}{\text{DCiE}} = \frac{\text{Total Facility Energy}}{\text{ITE Energy}} \quad (5)$$

where total facility energy includes power used by chiller, pumps, blowers, fans, and lighting. A direct measurement of pump power associated with the ES2 data center lab was not feasible because the chiller pump is shared with the building. However, the power consumed by pumps is negligible compared to chiller and blower power, and therefore is not considered in the calculations. Table 6 shows a 6% improvement in PUE in case 2 compared to case 1. Recently, mechanical load component (MLC) was introduced by ASHRAE as an improved measure for the mechanical efficiency in data centers. The MLC is defined as the sum of all

cooling, fan, pump, and heat rejection power divided by the data center ITE power. Following the above assumptions, MLC is calculated for both cases which shows 18% improvement in case 2.

Table 6: PUE and MLC analysis

Case #	Blower Power (kW)	Chiller Power (kW)	IT power (kW)	Lighting (kW)	PUE	MLC
1	8.1	14.9	44.8	0.2	1.52	0.51
2	3.4	9.5	30.4	0.2	1.43	0.42

7. Conclusions and Next Steps

In recent years, controlled airflow management has gained significant interest in the data center industry to mitigate increasing energy costs while ensuring reliability and availability. According to various surveys, IT equipment is the most significant power consumer in data centers followed by chiller plants and air handling units. An intelligent and CPU aware load balancing scheme can consolidate the IT load into a smaller number of servers and put the non-utilized servers into sleep or powered off mode. This can significantly decrease IT power consumption and chiller power. However, airflow delivery to the aisle is often not correspondingly adjusted in the data centers which can cause over/under-provisioning in some of the aisles.

In this paper, remotely controllable air dampers are utilized to ensure just enough air is supplied to an aisle at different airflow demands due to variations in IT load. The differential pressure between the CAC and room is considered as the control parameter. This allows designing a control system which is independent of the model, generation, layout, workload and the number of installed ITE and also eliminates the need for characterization of the servers. A fuzzy controller is designed to regulate the OAR of dampers based on the pressure measurements in the CAC. This controller ensures just enough airflow is delivered to the aisle with a small over-provisioning margin. The performance of the controller is tested for various load balancing schemes. The designed controller successfully regulated CAC pressure in various cases by adjusting supplied air flow to the aisle via controlling the OAR of dampers, and thus ensured proper provisioning of ITE in both decreasing and increasing IT load scenarios. It is shown that powering off the servers near the directional tiles (at the bottom half of the racks) can increase cold air bypass. Also, it is shown that utilizing a load balancer without assuring proper provisioning of an aisle can lead to recirculation through the servers and create hot spots. It is demonstrated that implementing the proposed airflow management scheme along with a smart load balancer in aisle C of the ES2 Data Center Lab can save up to 58% and 36% in the CRAH blower and the chiller powers, respectively.

In this study, the controller is tested in aisle C of the ES2 Data Center Lab successfully. However, a potential challenge can be the implementation of the controller in multiple aisles which are connected through a common plenum. In a multi-aisle data center with a shared raised floor, adjusting the OAR of dampers in one aisle can affect air delivery to other aisles. As a result, the OAR of dampers in other aisles need to intelligently vary to respond to this change. However, this can cause a cascading change in OARs which increases the overall response time of the control system. An important parameter in such airflow management systems is the ability to maintain a fixed plenum pressure by adjusting the blower speed of the cooling units and allowing sufficient tolerance for the ideal pressure range. The next step would be designing a holistic control approach which manages blower speed of the cooling units based on various pertinent parameters including IT load, number of active servers, plenum pressure and other factors.

Acknowledgments

The authors would like to thank U. L. N. Puvvadi and K. D. Hall from Binghamton University Computer Science and Data Center Group. We would also like to thank Russ Tipton and Arash Golafshan from Vertiv, and Mark Seymour from Future Facilities for their support and advice. This work is supported by NSF IUCRC Award No. IIP-1738793 and MRI Award No. CNS1040666.

References

[1] Sundaralingam, V., Arghode, V. K., Joshi, Y., and Phelps, W., 2014, "Experimental Characterization of Various Cold Aisle Containment Configurations for Data Centers," J. Electron. Packag., 137(1), p. 11007.

[2] Nemati, K., Alissa, H. A., Murray, B. T., and Sammakia, B., 2016, "Steady-State and Transient Comparison of Cold and Hot Aisle Containment and Chimney," *In Thermal and Thermomechanical Phenomena in Electronic Systems (ITherm), 15th IEEE Intersociety Conference On*, pp. 1435–1443.

[3] Bharath, M., Shrivastava, S. K., Ibrahim, M., Alkharabsheh, S. A., and Sammakia, B. G., 2013, "Impact of Cold Aisle Containment on Thermal Performance of Data Center," *InterPACK2013-73201*, pp. 1–5.

[4] Niemann, J., Brown, K., and Avelar, V., 2011, "Impact of Hot and Cold Aisle Containment on Data Center Temperature and Efficiency," Schneider Electr. Data Cent. Sci. Center, White Pap., 135, pp. 1–14.

[5] Alissa, H. A., Nemati, K., Sammakia, B. G., Schneebeli, K., Schmidt, R. R., and Seymour, M. J., 2016, "Chip to Facility Ramifications of Containment Solution on IT Airflow and Uptime," IEEE Trans. Components, Packag. Manuf. Technol., 6(1), pp. 67–78.

[6] Khalili, S., Alissa, H., Desu, A., Sammakia, B., and Ghose, K., 2018, "An Experimental Analysis of Hot Aisle Containment Systems," *Proceedings of the 17th InterSociety Conference on Thermal and Thermomechanical Phenomena in Electronic Systems, ITherm 2018*, San Diego, CA USA, pp. 748–760.

[7] Makwana, Y. U., Calder, A. R., and Shrivastava, S. K., 2014, "Benefits of Properly Sealing a Cold Aisle Containment System," Thermomechanical Phenom. Electron. Syst. - Proceedings Intersoc. Conf. ITherm2014, pp. 793–797.

[8] Patterson, M. K., Weidmann, R., Leberecht, M., Mair, M., and Libby, R. M., 2011, "An Investigation Into Cooling System Control Strategies for Data Center Airflow Containment Architectures," pp. 479–488.

[9] Shrivastava, S. K., and Ibrahim, M., 2013, "Benefit of Cold Aisle Containment During Cooling Failure," ASME 2013 Int. Tech. Conf. Exhib. Packag. Integr. Electron. Photonic Microsystems, (55768), p. V002T09A021.

[10] Alissa, H. A., Nemati, K., Sammakia, B. G., Seymour, M. J., Tipton, R., Mendo, D., Demetriou, D. W., Schneebeli, K., Tipton, R., Mendo, D., Demetriou, D. W., and Schneebeli, K., 2016, "Chip to Chiller Experimental Cooling Failure Analysis of Data Centers: The Interaction Between IT and Facility," IEEE Trans.

Components, Packag. Manuf. Technol., 6(9), pp. 1361–1378.

[11] Khalili, S., Tradat, M., Nemati, K., Seymour, M., and Sammakia, B., 2018, "Impact of Tile Design on the Thermal Performance of Open and Enclosed Aisles," J. Electron. Packag., 140(1), pp. 010907-010907-12.

[12] Patterson, M. K., Weidmann, R., Leberecht, M., Mair, M., and Libby, R. M., 2011, "An Investigation Into Cooling System Control Strategies for Data Center Airflow Containment Architectures," (44625), pp. 479–488.

[13] Gmach, D., Rolia, J., Cherkasova, L., and Kemper, A., 2009, "Resource Pool Management: Reactive versus Proactive or Let's Be Friends," Comput. Networks, 53(17), pp. 2905–2922.

[14] Arlitt, M., and Jin, T., 2000, "A Workload Characterization Study of the 1998 World Cup Web Site," IEEE Netw., 14(3), pp. 30–37.

[15] Stachecki, T. J., and Ghose, K., 2015, "Short-Term Load Prediction and Energy-Aware Load Balancing for Data Centers Serving Online Requests*."

[16] Gandhi, A., 2013, "Dynamic Server Provisioning for Data Center Power Management," PhD diss., Intel, (June), pp. 1–174.

[17] Atikoglu, B., Xu, Y., Frachtenberg, E., Jiang, S., and Paleczny, M., 2012, "Workload Analysis of a Large-Scale Key-Value Store," *ACM SIGMETRICS Performance Evaluation Review*, ACM, pp. 53–64.

[18] Hazelwood, K., Bird, S., Brooks, D., Chintala, S., Diril, U., Dzhulgakov, D., Fawzy, M., Jia, B., Jia, Y., Kalro, A., Law, J., Lee, K., Lu, J., Noordhuis, P., Smelyanskiy, M., Xiong, L., and Wang, X., 2018, "Applied Machine Learning at Facebook: A Datacenter Infrastructure Perspective," *2018 IEEE International Symposium on High Performance Computer Architecture (HPCA)*, pp. 620–629.

[19] Arghode, V. K., Sundaralingam, V., and Joshi, Y., 2016, "Airflow Management in a Contained Cold Aisle Using Active Fan Tiles for Energy Efficient Data- Center Operation Airflow Management in a Contained Cold Aisle Using Active Fan Tiles for Energy Efficient Data-Center," 7632(October 2017).

[20] Alissa, H. A., Nemati, K., Sammakia, B., Ghose, K., Seymour, M., and Schmidt, R., 2015, "Innovative Approaches of Experimentally Guided CFD Modeling for Data Centers," Therm. Meas. Model. Manag. Symp. (SEMI-THERM), 2015 31st, pp. 176–184.

[21] Chen, K., Federspiel, C. C., Auslander, D. M., Bash, C. E., and Patel, C. D., 2006, "Control Strategies for Plenum Optimization in Raised Floor Data Centers," HP Lab. White Pap.

[22] Khalili, S., Alissa, H., Nemati, K., Seymour, M., Curtis, R., Moss, D., and Sammakia, B., 2018, "Impact of Internal Design on the Efficiency Of IT Equipment In A Hot Aisle Containment System - An Experimental Study," *ASME 2018 IPACK*, San Francisco, CA, p. V001T02A009.

[23] Zadeh, L. A., 1973, "Outline of a New Approach to the Analysis of Complex Systems and Decision Processes," Syst. Man Cybern. IEEE Trans., (1), pp. 28–44.

[24] Athavale, J., Joshi, Y., Yoda, M., and Phelps, W., "Impact of Active Tiles on Data Center Flow and Temperature Distribution."

[25] Alissa, H. A., Nemati, K., Puvvadi, U. L. N., Sammakia, B. G., Schneebeli, K., Seymour, M., and Gregory, T., 2016, "Analysis of Airflow Imbalances in an Open Compute High Density Storage Data Center," Appl. Therm. Eng., 108, pp. 937–950.

[26] Alkharabsheh, S. A., Sammakia, B. G., and Shrivastava, S. K., 2015, "Experimentally Validated Computational Fluid Dynamics Model for a Data Center With Cold Aisle Containment," J. Electron. Packag., 137(2), pp. 21010–21019.

[27] Breen, T. J., Walsh, E. J., Bash, C. E., Punch, J., Shah, A. J., Bash, C. E., Rubenstein, B., Heath, S., and Kumari, N., 2011,

"From Chip to Cooling Tower Data Center Modeling: Influence of Air-Stream Containment on Operating Efficiency," *ASME/JSME Thermal Engineering Joint Conference*, pp. T10074-T10074-10.

[28] Iyengar, M., and Schmidt, R., 2009, "Analytical Modeling for Thermodynamic Characterization of Data Center Cooling Systems," J. Electron. Packag., **131**(2), p. 021009.

[29] Rubenstein, B. A., Zeighami, R., Lankston, R., and Peterson, E., 2010, "Hybrid Cooled Data Center Using above Ambient Liquid Cooling," *2010 12th IEEE Intersociety Conference on Thermal and Thermomechanical Phenomena in Electronic Systems*, pp. 1–10.

[30] Cengel, Y. A., and Boles, M. A., 1994, *Thermodynamics: An Engineering Approach*, New York.

[31] Breen, T. J., Walsh, E. J., Punch, J., Shah, A. J., and Bash, C. E., 2010, "From Chip to Cooling Tower Data Center Modeling: Part I Influence of Server Inlet Temperature and Temperature Rise across Cabinet," *2010 12th IEEE Intersociety Conference on Thermal and Thermomechanical Phenomena in Electronic Systems*, pp. 1–10.

IEEE
445 Hoes Lane
Piscataway, NJ 08854-4141

ISBN 978-1-7281-9129-4